Materials Science
in
Microelectronics
Volume II

Elsevier Internet Homepage
http://www.elsevier.com

Consult the Elsevier homepage for full catalogue information on all books, journals and electronic products and services.

Elsevier Titles of Related Interest

MATERIALS SCIENCE IN MICROELECTRONICS I:
The Relationship between Thin Film Processing and Structure
Eugene Machlin
2005, ISBN: 0080 44640 X
www.elsevier.com/locate/isbn/008044640X

HANDBOOK OF THIN FILMS
Five-Volume Set, 1–5
Hari Nalwa
2002, ISBN: 0125 12908 4
www.elsevier.com/locate/isbn/0125129084

NANOPHOTONICS:
Integrating Photochemistry, Optics and Nano/Bio Materials Studies, 1
Hiroshi Masuhara
2004, ISBN: 0444 51765 0
www.elsevier.com/locate/isbn/0444517650

Related Journals
The following journals related to materials science of thin films can all be found at
http://www.sciencedirect.com

Colloids and Surfaces A: Physicochemical and Engineering Aspects
Journal of Crystal Growth
Materials Science and Engineering: B
Surface and Coatings Technology
Progress in Organic Coatings
Materials Science and Engineering: A
Physics Reports
Thermochimica Acta

To Contact the Publisher
Elsevier welcomes enquiries concerning publishing proposals: books, journal special issues, conference proceedings, etc. All formats and media can be considered. Should you have a publishing proposal you wish to discuss, please contact, without obligation, the commissioning editor responsible for Elsevier's materials science books publishing programme:

David Sleeman
Commissioning Editor, Materials Science and Engineering
Elsevier Limited
The Boulevard, Langford Lane   Tel.: +44 1865 84 3265
Kidlington, Oxford   Fax: +44 1865 84 3929
OX5 1GB, UK   E-mail: d.sleeman@elsevier.com

General enquiries including placing orders, should be directed to Elsevier's Regional Sales Offices – please access the Elsevier internet homepage for full contact details.

Materials Science in Microelectronics Volume II

The effects of structure on properties in thin films

E.S. MACHLIN

Columbia University, New York, USA

ELSEVIER

AMSTERDAM • BOSTON • HEIDELBERG • LONDON • NEW YORK • OXFORD
PARIS • SAN DIEGO • SAN FRANCISCO • SINGAPORE • SYDNEY • TOKYO

ELSEVIER B.V.
Radarweg 29
P.O. Box 211,
1000 AE Amsterdam
The Netherlands

ELSEVIER Inc.
525 B Street, Suite 1900
San Diego,
CA 92101-4495
USA

ELSEVIER Ltd
The Boulevard, Langford Lane
Kidlington, Oxford
OX5 1GB
UK

ELSEVIER Ltd
84 Theobalds Road
London
WC1X 8RR
UK

First edition 1998

Second edition 2006

British Library Cataloguing in Publication Data
A catalogue record for this book is available from the British Library

Library of Congress Cataloguing in Publication Data
A catalogue record for this book is available from the Library of Congress

ISBN-13: 978-0-08-044639-4
ISBN-10: 0-08-044639-6

For information on all Elsevier publications visit our website at
www.books.elsevier.com

Typeset by Charon Tec Pvt. Ltd, Chennai, India
www.charontec.com
Printed and bound in Great Britain

Working together to grow
libraries in developing countries

www.elsevier.com | www.bookaid.org | www.sabre.org

ELSEVIER   BOOK AID
International   Sabre Foundation

Contents

Chapter III

Optical Properties 103

Chapter IV
Mechanical Properties 135

Chapter VII
Defects and Properties 215

Acknowledgement

My grateful thanks are due to Ujwal Ranadive, Librarian, Engineering Library, Columbia University who faxed me those articles I was unable to obtain via the Internet. Further, my thanks to Columbia University for maintaining my library privileges which enabled me to access publications on the Web without having to pay to read them. (I feel strongly that access to information should not be hindered by a pay to read barrier, a procedure that will return our civilization to the dark ages.) My thanks to Google Scholar which made my searches for information more efficient relative to Google. Thanks are due to David Sleeman, my editor at Elsevier, for his unstinting aid in unravelling problems. Last but not least my thanks to Edda for being.

CHAPTER I

Electrical Properties

Various electrical properties of thin films are vital to the efficient functioning of many devices. For example: interconnections in integrated circuits require low resistivity; thin-film transistors require adequate charge carrier mobility and high on–off current ratio; solar cells require a high value of the minority carrier diffusion length, a low dark conductivity coupled with a high photoconductivity, and as high a short-circuit current and open circuit voltage as possible; superconducting thin films require a high critical current density; dielectric films require a high breakdown voltage; ohmic contacts require a low interfacial impedance; etc. *We will consider these electrical properties and how the structure (i.e. mostly defect structure) affects them in this chapter. As stated in the Preface, we shall not give a detailed exposition of each topic. Rather, our objective here is to provide, where possible, an understanding of the physical bases for the effects of structure on the electrical properties and a summary of the data that characterize these effects.*

1. Conductivity and charge carrier mobility.

1.1. Metallic conductors.

1.1.1. Conducting lines (interconnections) in integrated circuits.

1.1.1.1. History.

The lower is the electrical resistance of the interconnection material in integrated circuits the faster can signals be transmitted between devices. Hence, there is a compelling commercial reason to make use of an interconnection material having the lowest electrical resistivity. Yet, the standard interconnection material in use prior to 2002 in integrated circuits is an aluminum alloy having somewhat higher resistivity than pure aluminum, and even higher resistivity than other metals such as copper, gold and silver. (Table 1.1 provides data which support this statement.) Why?

Table 1.1. Electrical resistivity of metals*

Metal	Electrical resistivity (20 °C) $\mu\Omega$-cm
Al	2.65
Cu	1.67
Au	2.35
Ag	1.59
W	5.65

* Data taken from Metals Handbook, 8th edition, ASM, Metals Park, Ohio.

Silver is not used as an interconnect material because it is too prone to attack by sulfur and oxygen and cannot be relied upon to maintain its low resistivity value for the lifetime of a computer. Copper is also prone to oxidation, albeit, not as readily as silver. Further, at the time that thin film interconnections were first used, the vacuum maintained in the deposition chambers was insufficient to prevent a high oxygen concentration from being incorporated into solid solution in the deposited copper films with an attendant excess resistivity due to the impurity content. The oxygen solubility in aluminum is much lower than in copper and aluminum develops a thin protective oxide that prevents further oxidation, whereas copper does not. Thus, aluminum was chosen as the standard interconnection material. Subsequent problems with electromigration induced failures of interconnection lines of aluminum brought about the use of aluminum alloys of still higher resistivity. With the trend towards narrower interconnections and concomitant increase in the interconnection resistance (decrease in signal propagation velocity) the industry has now begun to use copper based interconnection technology.*

1.1.1.2. Thin film versus bulk resistivity.

The question arises as to whether the resistivity of a thin film is the same as that of the bulk material. In the answer to this question we encounter the effect of thin film structure and dimensions on electrical resistivity. *Generally, as-deposited films have higher values of resistivity than their corresponding bulk metallic form.* Post-deposition annealing of the films usually produces an approach of the resistivity values to those characteristic of bulk materials in thick films (>1 μm). A reasonable deduction based on the foregoing generalization is that defects which affect the resistivity are annealed out in the annealing step. Are there data which can determine the validity of this deduction? The answer is in the affirmative and details are given below.

It is useful at this point to recall from theory that we should expect that defects that displace atoms from the sites of a perfect periodic lattice act to scatter electrons and affect the room temperature resistivity significantly when the distance between defects becomes on the order of or less than the mean free path between thermal scattering events for electrons in single crystals at this temperature. Since the latter quantity for the metals under consideration varies from about 400 to 600 Å then we may expect defects to affect the room temperature resistivity when the distance between them is on the order of or less than this mean free path length.

The annealable defects that may contribute to the excess resistivity in as-deposited films are point defects, dislocations, stacking faults, grain and twin boundaries, and non-soluble impurity atoms. The contributions of each of these potential sources of excess resistivity may be evaluated. For example, the increment

---

* The problems with Cu interconnects are discussed in Chapter V.

in resistivity for vacancies or self interstitials in these metals is on the order of 1 $\mu\Omega$-cm/at%.[1a] Since there has been one observation that suggests that about 1 at% vacancies might be present in a gold film deposited at room temperature by physical vapor deposition,[1b] it is reasonable to suspect that vacancies in high melting point metals, in which vacancies are not likely to anneal out rapidly at room temperature, may contribute to the annealable excess of the resistivity in these metal films. (The excess resistivity in Cu due to deposited vacancies falls to a zero value above about 150°K.[1c]) Self interstitials should make no contribution to the excess resistivity in metal films because their low activation energy for migration in metals[1a] assures their disappearance in such room temperature deposited films.

The dislocation density in silver deposited at room temperature in the absence of energetic particle bombardment was reported[2a] to be somewhat less than 10^{11} cm^{-2} and only increased to above a density of 10^{12} cm^{-2} when the average energy per depositing atom exceeded about 40 eV. Measurement of the excess resistivity due to dislocations in cold-worked copper[2b] yielded only a 2% increment in the resistivity (<0.02 $\mu\Omega$-cm) for an increment in dislocation density of $4 \cdot 10^{12}$ cm^{-2}. Thus, there is no basis for assigning to dislocations the excess in the resistivity of films relative to that of bulk metals, at least in films deposited in the absence of energetic particle bombardment.

A significant observation has been made by De Vries[3] who noted that in many metal films the excess resistivity is inversely proportional to the grain size and the film thickness. That the as-deposited average grain size in physical vapor deposited (PVD) films is proportional to the film thickness and, indeed, approximately equal to it is well known (e.g. see Ref. [4]). De Vries[3,4] was able to separate out the contribution of surface scattering of electrons from that of grain boundary scattering by measuring the derivative of the excess resistivity with respect to temperature. From theory[5], this derivative for surface scattering should deviate from the corresponding derivative of the bulk resistivity, whereas it should equal that for bulk resistivity for the case of grain boundary scattering. He found that for all the metal films studied, the temperature derivative for the excess resistivity equalled that for the bulk resistivity. Hence, the excess scattering could not stem from surface scattering. Further, since the excess resistivity was found to depend inversely on the grain size, and since neither the as-deposited excess vacancy concentration, impurity concentration, nor the dislocation density would be expected to vary with the grain size, then the conclusion may be drawn that the excess resistivity in films produced by PVD is due to grain boundary scattering.

Additional observations support this conclusion. For example, the dislocation density can be increased by a factor of ten in deposition using energetic particles with the dislocation density a monotonic increasing function of the normalized energy per particle.[2a] Ziemann and Kay[6] have found, however, that the residual resistivity of films produced via sputtering with simultaneous ion bombardment is not a monotonic function of the normalized energy. Further, their data reveal about the same value of the constant of proportionality between the excess resistivity and the inverse grain size as that of de Vries[3] for the same metal, Pd. These results are consistent with our previous conclusion that dislocations are not responsible for the excess resistivity of as-deposited metal films.

Actually, according to theory the excess resistivity should be proportional to the inverse grain size only for values of the grain size below about 1 micron (1 μm). This is apparent upon substitution of the appropriate values in the following equation due to Mayadas et al.[7] that describes the effect of grain boundary scattering upon the resistivity:

$$\rho_{gr} = \rho_\infty/[1 - 1.5\alpha + 3\alpha^2 - 3\alpha^3 \ln(1 + \alpha^{-1})] \qquad (1.0)$$

and

$$\alpha = (\lambda_\infty/D)[R/(1 - R)]$$

where λ_∞ is the electron mean free path in single crystal bulk material, D is the average in-plane grain diameter, and R is the reflection coefficient of the conduction electrons striking the grain boundaries. (Measured values of R range between 0.17 and 0.25.) As shown, in Volume I of this series* (p. 63) the as-deposited[125] grain size in metallic films produced by PVD is less than about 0.1 micron. Substituting, D = 0.1 μm, λ_∞ = 0.04 μm (Cu), R = 0.2 into equation (1.0) yields ρ_{gr}/ρ_∞ = 1.15. Hence, grain boundary scattering contributes a significant excess resistivity in as-deposited films of low resistivity metals that must be removed by annealing in order for the interconnections to function at their highest efficiency. The same conclusion probably applies to other metallic conducting interconnection materials, such as titanium and cobalt disilicides, i.e. the excess of the resistivity of an as-deposited metallic conducting thin film over that of the bulk material is likely to be due to grain boundary scattering.

When films become thinner than the electron mean free path it becomes possible to have a surface scattering contribution to the resistivity. Structure affects the magnitude of this contribution and is manifested as an effect of the roughness at the interfaces of the film. When the scale of the roughness (i.e. the distance between peaks) exceeds the deBroglie wave length of the electrons then surface scattering in such thin films becomes significant and its contribution to the resistivity is also significant. For a roughness that prevents any specular reflection of electrons from the interfaces, the dependence of resistivity on film thickness is given by the relation $\rho = \rho_0/(1 - 3\lambda_0/8d)$, where λ_0 is the electron mean free path length and d is the film thickness. However, when grain size is much smaller than the film thickness (i.e. as-deposited grain sizes are 100 Å or smaller) then the effect of grain boundary scattering on resistivity should overwhelm that of the surface scattering.

* E.S. Machlin, in **Materials Science in Microelectronics. I. The Relationships Between Thin Film Processing and Structure**, Elsevier, Oxford, UK, 2005.

1.1.2. Contacts.

Contact materials between metallic interconnections and semiconductors reveal at least one additional direct effect of structure on resistivity – the effect of crystal structure on resistivity. This effect is manifested in two silicides that are used as contact materials: $CoSi_2$ and $TiSi_2$. Several metastable crystal structures with different resistivities at the $CoSi_2$ composition have been found in thin films. The desired low resistivity structure is the CaF_2 structure. A CsCl structure having vacancies on the Co type site and an adiamantine structure are the metastable structures that appear. The template mode of growth* of $CoSi_2$ favors the formation of the defected CsCl structure.[98] The adiamantine structure appeared in the formation of $CoSi_2$ when formed from Co/Ti bilayers.[99] The pinholes that have been found in $CoSi_2$ films produced by the template method probably derive from transformation of the defected CsCl structure to the more stable CaF_2 structure. These metastable structures are resistant to transformation to the stable structure and there may be untransformed grains after a heat treatment carried out to induce the transformation. It has been suggested that the poorer conductivity found in the template method of forming $CoSi_2$ films is due to the presence of such untransformed regions. Polycrystalline $CoSi_2$ (CaF_2 type) has a resistivity of 15–18 $\mu\Omega$-cm, whereas the monocrystalline, epitaxial film has a resistivity of about 10 $\mu\Omega$-cm. Thus, grain boundary scattering in the former is a possible source of enhanced resistivity relative to that of the epitaxial film.

Up until recently the C-54 structure of $TiSi_2$ was produced by solid state transformation from the higher resistivity C-49 structure by annealing at a high temperature. In this normal annealing procedure of converting the C-49 to the C-54 structure it is found that the surface roughness increases with the annealing temperature, a result that is detrimental to properties. The resistivity of the C-54 allotrope is about 15 $\mu\Omega$-cm. It has now been found possible to produce the C-54 structure at a lower temperature by sputtering Ti onto a substrate of Si, the surface layer of which has been previously made amorphous, held at about 450°C and then given a rapid thermal anneal[100a] and also by alloying the Ti with Mo, Nb, or Ta.[100b]

Another property a contact material must possess is that of thermodynamic stability with respect to chemical reaction with the materials it contacts. Silicide type contacts to silicon generally have this property, but metallic contacts to compound semiconductors generally are not thermodynamically stable and often produce products of high contact resistance. The effect of structure on the resistance to chemical reaction between contact material and its contacting materials is considered in Chapter V.

A possible equivalent to a silicide for contact to germanium is Cu_3Ge, which has a long-range-ordered monoclinic structure. This germanide has the

---

* See Volume I, p. 198 of this series for a description of this method.

remarkable low resistivity of 6 $\mu\Omega$-cm at room temperature.[112] The electron mean free path is about 1200 Å in a film of grain size varying from 1000 to 5000 Å and the main contribution to the resistivity in films less than about 1000 Å thick is from surface scattering.

1.2. Charge carrier mobility in semiconductors.

An important figure of merit in semiconductor devices, such as a thin film transistor (TFT), is the mobility, defined as the charge carrier velocity per unit field strength. The higher is the mobility, the less is the time required for the device to perform an operation. There are several potential limitations to the value that the mobility can attain in semiconductors. Acoustic and optical phonons can scatter charge carriers. Such scattering determines the maximum value for the mobility in the absence of all other sources of charge carrier scattering. (Any text on semiconductor or solid state physics provides a detailed description of the theory of such thermal scattering.) The other significant sources of scattering in monocrystalline, inorganic semiconductors are extrinsic and arise from ionized impurity atoms and charged defects and will be considered below.

1.2.1. Effect of ionized impurity atoms on charge carrier mobility in monocrystalline semiconductors.

The electrical resistivity, ρ, in semiconductors equals the inverse of the conductivity, σ, and the latter equals:

$$\sigma = e(n\mu_n + p\mu_p) \tag{1.1}$$

where n and p (for negative carriers and positive carriers) are the densities of electrons and holes, respectively, and μ_n and μ_p their respective mobilities. Thus, the structural dependence of the conductivity or resistivity, aside from that achieved through the addition of dopant atoms to control n or p, would be manifested, if at all, through structural effects on the mobilities. One such structural effect is that due to impurity atoms. The theory describing the effect of charge carrier scattering from ionized impurities on mobility in non-polar semiconductors has been developed by Conwell and Weisskopf.[8] Their result for the mobility resulting from such scattering is:

$$\mu_i = C\kappa^2 T^{3/2} N_I^{-1} m^{*-1/2} \{\ln[1 + x^2]\}^{-1} \tag{1.2}$$

where

$$x = C^* T\kappa N_I^{-1/3} \tag{1.3}$$

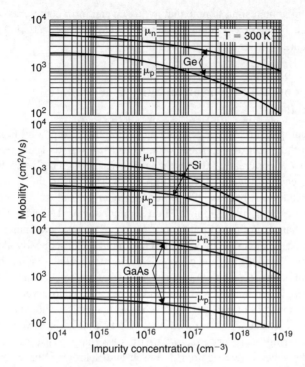

Figure 1.1. Effect of impurity concentration on mobility. (S.M. Sze, in **Physics of Semiconductor Devices**. Copyright 1981, reprinted with permission of John Wiley & Sons, Inc.)

and T is the absolute temperature, N_I is the density of ionizable impurities, κ is the permittivity of the semiconductor, m* is the effective mass of the charge carrier, and C and C* are numerical constants.

The effective charge carrier mobility, μ, due to both impurity scattering, μ_I, and to thermal scattering, μ_T, is given by:

$$\mu^{-1} = \mu_I^{-1} + \mu_T^{-1} \tag{1.4}$$

Figure 1.1 shows how the experimental drift mobility at room temperature varies with impurity concentration in three semiconductors. As shown, for impurity concentrations below about 10^{15} per cm^3, the observed mobilities for both holes and electrons are independent of the impurity concentration and are thus due to thermal scattering alone. With impurity densities increasing above 10^{15} per cm^3 the observed mobilities decrease, as expected from the C–W relationship. Also, the observed mobility varies inversely with the effective mass of the charge carrier, in agreement with both the C–W relation and theory for thermal scattering by phonons.[8,9]

Incidentally, it is interesting to note that hydrogen can form complexes with certain dopants which, besides passivating the dopants, act to produce scattering centers that decrease minority carrier mobility in certain cases.[109]

1.2.2. Effect of dislocations on mobility in crystalline semiconductors.

A comparison of the effect of ionized impurities on mobility in semiconductors with the effect of impurities on the electrical resistivity in metals reveals that the former effect greatly exceeds the latter one although both effects are due to scattering of charge wave packets by impurities. The reason the former effect is larger in magnitude is because the impurity is charged and, hence, the scattering cross-section of the ionized impurity in a semiconductor (typically $10^{-12}\,cm^2$) is much larger than that of the impurity in a metal (i.e. charge screening is much less effective in a semiconductor as compared to a metal). Consequently, the effects of charged defects on mobility (and conductivity) in semiconductors may be expected to be larger than the corresponding effects of uncharged defects on conductivity in metals. Indeed, in semiconductors an effect on mobility is observed[12a,106,118,119,122] at dislocation densities of 10^7–$10^8\,cm^{-2}$, whereas in metals, as noted previously, the density must exceed about $10^{12}\,cm^{-2}$ for an effect on the resistivity to be observed.

Scattering of majority charge carriers by charged dislocations has been considered by several investigators. Labusch and Schröter[12a] have reviewed this work, at least those developed prior to 1980. According to these theories the reciprocal mobility depends upon dislocation density, N_D, and temperature, T, by:

$$\mu_D^{-1} = (N_D/T)[A + B\lambda Q_D^2/T^{1/2}]$$

where λ is the screening length and Q_D is the charge per unit length of dislocation.

This inverse dependence of the mobility on dislocation density has been found[106] for the Hall electron mobility in InSb at 77 K where the contribution of thermal scattering to the inverse mobility is not significant (see Figure 1.2(a)). Also, the decrease in mobility with decreasing temperature due to dislocation scattering has been observed by several investigators.[106,122]

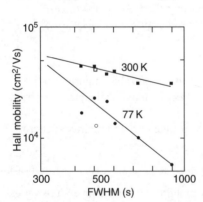

Figure 1.2. (a) Electron Hall mobilities as a function of rocking-curve width. The latter varies as the square root of the dislocation density. FWHM: full width at half-maximum. (From B.-S Yoo et al., MRS Symp. Proc. 325, 279(1994) with permission.)

These data are consistent with the expectation that charged dislocations should have a greater effect on the scattering of electrons than ionized impurities in that with on the order of 10^6 charges per cm of dislocation line these dislocation densities are still smaller than the concentration indicated in Figure 1.1 for the onset of an effect of ionized impurities on mobility. Incidentally, hydrogen passivates the detrimental effect of dislocations on the electron mobility and in films formed in the presence of hydrogen the value of the threshold dislocation density may appear to be higher than it would be in the absence of hydrogen. The associated charge carrier traps are impurity atoms, dangling bonds at jogs, kinks and other disconti-nuities along or near dislocation cores. This subject will be discussed in greater detail in Chapter VII. Another effect of structure is associated with the morphology of the dislocation lines. If sufficient unpassivated dislocations are present along a preferred orientation then they produce an anisotropy in the sheet resistivity.[108]

The electrical activity of stacking faults has been determined to be located at the partial dislocations that bound them.[124] A more detailed discussion of stacking faults is provided in Chapter VII.

1.2.3. Effect of defects on mobility of polycrystalline semiconductors.

1.2.3.1. Polycrystalline silicon: grain boundary versus intracrystalline defects.

The dependence of the resistivity of polycrystalline silicon on doping concentration differs from that for single crystal silicon as shown in Figure 1.2(b).

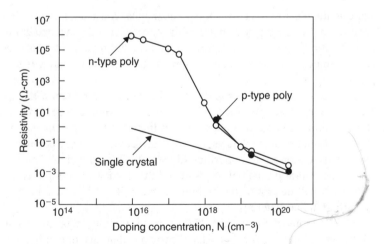

Figure 1.2. (b) Resistivity of poly-Si compared to that of single crystal Si. (From M.Y. Ghannam, in **Advanced Silicon and Semiconducting Silicon Alloy Based Materials and Devices**, eds J.F.A. Nils et al., IOP Publishing, Bristol, 1994 with permission. Copyright 1994 IOP Publishing, Ltd.)

At doping concentrations less than about $1 \cdot 10^{19}/cm^3$ the resistivity of polycrystalline silicon tends to be much higher than that for single crystal silicon. One contribution to this increment in resistivity is due to carrier trapping at the grain boundaries, which acts to reduce the free carrier concentration within the grains. However, it has been shown that this effect cannot by itself account for the increment in resistivity. Another origin of enhanced resistivity is the development of a potential barrier at the grain boundaries resulting from the excess charge that becomes trapped there.[12b] This potential barrier to charge flow across the grain boundary acts to reduce the mobility of the majority charge carrier. Many investigators have contributed to the development of the theory of grain boundary limited mobility.[13] The theory of Seto[14] will be used here because it provides explicitly for the dependence of mobility on grain size.

In Seto's model[14] the effective mobility is limited by thermionic emission of charge from the barrier and is given by:

$$\mu_{eff} = e(2\pi kTm^*)^{-1/2} D \exp(-eV_b/kT) \tag{1.5}$$

where

$$V_b = eND^2/8\kappa \quad \text{when } DN \ll N_t \tag{1.6}$$

or

$$V_b = eN_t^2/8\kappa N \quad \text{when } DN \gg N_t \tag{1.7}$$

and e is the electronic charge, k is Boltzmann's constant, T is the absolute temperature, m^* is the effective mass of the carrier, D is the average grain diameter, N is the number of ionized impurity atoms (including dopant atoms) per cm^3, N_t is the number of grain boundary trapping states per cm^2, and κ is the dielectric permittivity.

Seto's theory has been shown to be in good agreement with experiment.[15] We shall now make use of it to discuss the effect of structure on the mobility of polycrystalline silicon. It is apparent that there are three parameters in equations (1.5)–(1.7) that relate to defects: D and N_t are grain boundary parameters and N is the density of ionizable impurities (dopants). The number N depends upon the device incorporating the poly-Si film and the uncontrolled impurity content. The value of N will determine the mobility of the bulk material within the grains (i.e. the intragranular material).

We must recognize that the reciprocal mobility measured for a polycrystalline semiconductor film consists of two contributions: one of intragranular origin and the other of intergranular origin. Thus, to evaluate the latter we must devise some means of separating the observed mobility into the inter- and intragranular components. Seto's formulation (equations (1.5)–(1.7)) and the knowledge that the observed mobility is related to its contributions from various independent sources

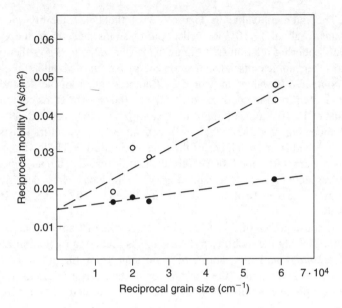

Figure 1.3. Plot of reciprocal mobility versus grain size. Open circles correspond to rapid thermally annealed films prior to hydrogenation. Closed circles to duplicate samples, but which have been subject to hydrogenation. (Data from Campo et al.[16])

via a relation of the form of equation (1.4) (i.e. $\mu_{obs}^{-1} = \Sigma_i \mu_i^{-1}$) provide the means to accomplish this objective, as we will now demonstrate.

An analysis of equation (1.6) indicates that a plot of the reciprocal of the mobility versus reciprocal average grain diameter should exhibit a minimum value at a grain diameter equal to $(4\kappa kT/e^2N)^{1/2}$ (which equals two Debye screening lengths). This plot should also yield an inflection point as the reciprocal average grain diameter increases and then should approach a straight line with the slope equal to $(2\pi kTm^*)^{1/2}/e$. Whereas in the regime corresponding to equation (1.7), the same plot should yield a straight line that has a positive value of the y-intercept.

We make use first of the results of Campo et al.[16] to illustrate the results expected for the regime defined by equation (1.7). Their data are plotted in Figure 1.3 in the form of $1/\mu$ versus $1/D$. (The field effect mobilities in this figure correspond to hole mobilities.) It is apparent that the lines through the points for a given treatment in Figure 1.3 do not extend through the origin. However, straight lines can be drawn through both sets of points that intersect the ordinate at one point. This result implies that the regime applicable to the data of Campo et al.[16] is that defined in equation (1.7). Thus, we suggest that the intragranular contribution to the observed reciprocal mobility is given by the intercept value in Figure 1.3 and that the intergranular contribution can be obtained from the difference between the reciprocal of the observed mobility and the intercept.

The intercept value in Figure 1.3 of the hole mobility ($66\,cm^2/Vs$) corresponds to about $1 \cdot 10^{19}$ ionized impurity atoms per cm^3 in the Conwell–Weisskopf scattering relation (see Figure 1.1). This value in Seto's theory corresponds to a thermionic emission activation energy barrier of about 0.022 eV, which may be compared to values measured by Campo et al from the dependence of mobility on reciprocal temperature of 0.027 and 0.07 eV for films containing the same value of $1 \cdot 10^{19}$ ionized impurities per cm^3, but which had different deposition and annealing procedures. Values of activation energy and trap density also may be evaluated from the slopes of the lines in Figure 1.3. These values of activation energy are 0.017 and 0.062 eV for the hydrogenated and non-hydrogenated samples, respectively. The corresponding grain boundary trap densities using $1 \cdot 10^{19}/cm^3$ as the density of ionized impurities are $3.0 \cdot 10^{12}$ and $5.73 \cdot 10^{12}/cm^2$, respectively.

A consequence of this analysis of the results of Campo et al. is that the main factor limiting the mobility in their films is scattering by the ionized impurity concentration and not thermionic emission from the grain boundaries! Incidentally, consideration of the contribution of the interior of grains to the limitation of the effective mobility is almost universally neglected in the poly-Si TFT literature.

Data in Ref. [17] can be used to illustrate the regime expected when equation (1.6) applies. These data are plotted in Figure 1.4 using the same coordinates as for Figure 1.3. The qualitative functional behavior of the data points is that expected from equation (1.5) and the regime of equation (1.6). We have used the experimental values of the grain diameter and mobility at the minimum in the curve in Figure 1.4 to determine a value of κ/N, which was then substituted into equations (1.5) and (1.6), to yield the curved line that lies below the data points. The vertical shift in μ^{-1} between the line and the data points increases as the grain diameter increases. We interpret this difference in mobility as due to intragrain origin. The difference in mobility between experiment and theory cannot be ascribed solely to scattering from ionized impurities within the grains. (The density of ionized impurities evaluated from the grain diameter corresponding to the minimum in μ^{-1} is $2.49 \cdot 10^{16}/cm^3$ and the value of μ_n^{-1} ($9 \cdot 10^{-4}\,Vs/cm^2$) corresponding to this density of ionized impurities is only one-fourth of the vertical shift between the data points and the calculated line ($4 \cdot 10^{-3}\,Vs/cm^2$).) Thus, we conclude that there is an intragrain contribution that reduces the effective mobility, that is not present in the very fine grains, and which increases as the grain size increases. We speculate that the defects that contribute to this intragrain barrier to the mobility are primarily stacking faults, which according to Kazmerski et al.[18] act as potential barriers to charge carrier flow in the same manner as grain boundaries. Incoherent microtwins act similarly.

Thus, in the regime defined by equation (1.6) the product ND^2 determines the barrier energy for emission of electronic charge from the grain boundaries and

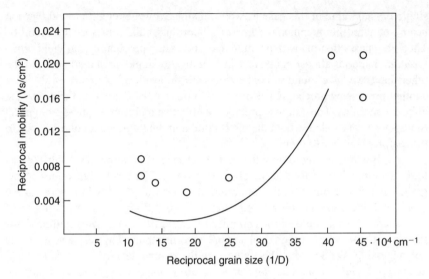

Figure 1.4. Plot of reciprocal mobility versus reciprocal grain size. (Data from Okumura et al.[17]) Full line corresponds to equation (1.6) fitted to have the minimum reciprocal mobility at the same grain size as that for the minimum reciprocal mobility of the experimental points.

the grain boundary limitation to the mobility. In addition, one may expect that intragrain defects can contribute to the mobility limitation, especially in grains larger than about $3 \cdot 10^{-6}$ cm in diameter. In this regime the mobility is independent of the density of grain boundary traps. However, in the regime defined by equation (1.7) this density affects the limitation to the reciprocal mobility due to grain boundaries.

Given the above results and theory one may wonder whether it is possible to produce defect-free regions within the grains. In this regard we may note that Baba et al.[19] have found, with their particular solid phase crystallization (SPC) procedure of producing large grain poly-Si TFT's with *the active region of the TFT inside one grain*, a mobility value nearly equal to that which would be obtained in equivalent single crystal silicon. In particular, the value found was a Hall electron mobility of 808 cm²/Vs, which is 80% of the corresponding single crystal value at the charge carrier concentration of about $1.5 \cdot 10^{16}$/cm³. Also, Sameshima et al.[20] reported, in films produced by a particular laser crystallization method, a field effect electron mobility of 600 cm²/Vs for a carrier concentration of $1.8 \cdot 10^{11}$/cm², which for an active channel 20 nm thick corresponds to a carrier concentration of $0.9 \cdot 10^{17}$/cm³. The electron mobility of single crystal silicon at this carrier concentration is about 750 cm²/Vs. Hence, the mobility obtained is about 80% of the single crystal value in this case as well. Finally, Schwebel et al.[21] found that ion beam sputtered homoepitaxial films of silicon had bulk values of the mobility. Thus, it appears that many deposition and processing procedures can yield *intra-granular* material almost equivalent in mobility to single crystal silicon. On the other hand, we have found in connection with the work of Okumura et al.[17] that another processing procedure has produced intragranular defects that diminish the effective mobility. Thus, in interpreting the effective mobility obtained from poly-Si it is necessary to know both the intergranular and intragranular contributions to the reciprocal effective mobility.

It is interesting to note that in the absence of intragranular contributions to the reciprocal mobility that the effective mobility can reach high values at low grain sizes. For example, at the minimum in reciprocal mobility in Figure 1.4 the mobility value corresponding to equation (1.6) is $635\,cm^2/Vs$ at a grain size of 0.05 microns! Thus the electron mobility of $350\,cm^2/Vs$ in polycrystalline films that have a grain size between 0.04 and 0.2 microns reported by Serikawa et al.[22] is not surprising. We note that this value is in the regime defined by equation (1.6) *where the mobility is independent of the grain boundary trap density*. Since we have made use of the maximum values obtained to date in these small grain size films and since most of the reported values are factors of ten smaller than these values it is apparent that intragrain defects contribute significantly to the effective reciprocal mobility for most of the small grain poly-Si produced as yet.

Processing procedures have reached the point at this writing of being capable of forming single crystal regions at defined positions along a poly-Si film of dimensions larger than the area occupied by a transistor.[23] Thus, the mobility achievable for a transistor so located is limited only by the defects present in the intragranular material. The analysis given above suggests that the present strategy of developing a processing procedure that eliminates grain boundaries in the active region of a transistor is not the only possible strategy to follow to approach single crystal mobility values in poly-Si films. This other strategy is to produce poly-Si films which have a ND^2 product at the minimum of the reciprocal mobility in the regime where equation (1.6) is applicable and which are processed so that intragrain defects do not contribute to the effective reciprocal mobility. (However, this strategy neglects the need for a high on–off current ratio for useful TFT operation, which may be difficult to satisfy in the regime associated with equation (1.6).) Indeed, for both strategies it is necessary that this intragrain term be negligible.

A detailed examination of the relations between processing and intra-grain defect structure in poly-Si is provided in Volume I of this series. However, they will be summarized below for the interested reader.

As-deposited poly-Si exhibits stacking faults in {111} oriented grains for substrate deposition temperatures below about 485°C and above the limiting maximum temperature (~350°C) below which amorphous Si is deposited. Most *as-deposited poly-Si grains contain incoherent microtwins, the density of such microtwins is greatest for {110} oriented grains.* *

Poly-Si produced from as-deposited amorphous Si will contain a high density of dislocations (>10^{10} per cm^2) when the temperature of crystallization is at or above 600°C.

Poly-Si formed by crystallization of amorphous Si that has been deposited in a vacuum less clean than that corresponding to ultra high vacuum conditions (<10^{-8} Torr) will contain sufficient oxygen to induce the formation of twins and stacking faults at the amorphous–crystalline interface that are frozen-in the crystalline region behind the advancing interface.

Laser melted and crystallized Si with a solidification front normal to the <111> axis will contain copious twins and stacking faults for liquid/solid interface velocities between 6 and 15 m/s and will be free of such defects for solidification front velocities less than 6 m/s.

Thus far in this section we have devoted attention to the methods that allow one to distinguish between the contributions of intergrain (grain boundaries) and intragrain contributions to the reciprocal mobility. We have also noted that proper processing procedures can yield monocrystalline (defect-free) quality intragrain material and also grain boundary free active regions. Thus, we have explored how mobility in poly-Si films can be maximized. We have not devoted attention to the nature of the defects in the grain boundaries that act to trap charge there.

The identity of the defects that trap charge at the grain boundaries is not known. However, this lack of knowledge has not acted as a barrier to proposals concerning their identity. One group believes that these defects have a structural origin and are dangling bonds.[24] Another group believes that the charge trapping defects are point defects of the host lattice rather than structural imperfections at the grain boundary.[25] Still another group believes that the major contribution to the density of recombination centers at grain boundaries stems from impurities segregated there.[26] Yet another group considers all the grain boundary states in the energy gap to be charge trapping states which are then treated as being uniformly distributed throughout the grain volume to arrive at their effect on charge transport properties.[27] The identity of the charge traps at grain boundaries will be explored in Chapter VII. In this chapter we will maintain the viewpoint of a materials scientist who is concerned with the control of processing in order to improve properties.

It is well known[28] from experience with amorphous silicon (a-Si) that dangling bonds interact with electrons and holes to form charged entities and that hydrogen can satisfy such dangling bonds so that they passivate (i.e. do not trap charge). However, as was noted above in the analysis of the results of Campo et al.,[16] hydrogenation does not necessarily passivate all the grain boundary charge

---

* As in the normal notation the {hkl} indices refer to hkl type planes that are parallel to the film plane.

trapping states. Indeed, in the latter case, a significant density of such grain boundary states remains after hydrogenation. Ionized impurity atoms segregated at grain boundaries represent one class of grain boundary species that may be resistant to hydrogen passivation[29] and also act to trap charge. The specific impurities responsible for the extrinsic grain boundary defects are still unknown despite a host of investigations designed to answer this question.* It should be noted that impurities are likely to be introduced during the processing of a TFT structure. In particular, metal atoms from the electrodes and oxygen from gate and substrate oxides as well as from the chamber atmosphere along with carbon and nitrogen have the intrinsic capability of diffusing along interfaces and grain boundaries over distances greater than 0.01 cm in poly-Si at processing temperatures above 600°C. Care must be taken to prevent this potential activation of grain boundary defect states, if possible. It should be noted again that the density of grain boundary defects is mainly of interest from the practical viewpoint of device application only when equation (1.7) defines the regime of the active region of the TFT.

First principles calculations of the properties of grain boundaries in silicon and germanium have been reviewed by Sutton.[30] He came to the conclusion that the absence of band tails in the results of such calculations implies that intrinsic grain boundary defect states do not contribute to the band tails responsible for charge carrier trapping. He is partial to the suggestion that the band tails are due to segregated dopant atoms. However, he does not rule out the possibility that these calculations are not the last word on the subject because the calculations are based on ideal boundaries devoid of steps, facet junctions, continuous curvature of the boundary plane, point defects, and dislocations run in from the adjoining crystals. For example, it is well-known that coherent twins in silicon are electrically inactive yet a slight misorientation of the boundary from the twinning plane yields electrical activity.[31] Indeed, later first-principles calculations on twist boundaries in silicon revealed the presence of band-tail and mid-gap states in the energy gap.[32] The controversy between those who believe that the grain boundary trapping states have an extrinsic origin and those who believe these states have an intrinsic origin will be considered in Chapter VII.

Summarizing, regardless of our ignorance concerning the identity of the defects that diminish the mobility in poly-Si, processing procedures exist that allow the attainment of near single crystal mobilities over TFT channel lengths in poly-Si. The objective at this stage is to achieve such a result with the least expensive processing procedure over large areas of cheap substrate materials.

The potential of poly-Si TFTs is being actively explored, not only for application to flat panel displays, but also as a low cost replacement of chip based

---

* More recent investigations using a new X-ray beam induced current technique identify Fe as the main impurity in cheap multicrystalline Si. (O.F. Vyvenko et al., J. Appl. Phys. 91, 3614(2002); J. Phys. Condens. Matter 14, 13079(2002).)

transistor technology for various products. Thus, the materials scientist can make a significant contribution to expedite the realization of these applications of poly-Si.

1.2.3.2. Polycrystalline compound semiconductors.

In elemental semiconductors we have seen that in single crystals the mobility of charge carriers may be affected mainly by phonon scattering and charged point, line, and area defects. These same sources of scattering exist in compound semiconductors with contributions to charged point defects from the spectrum of additional possible intrinsic defects in off-stoichiometry compositions.

First, let us consider maximum values of the charged carrier mobilities that have been reported in the literature. Bulk single crystal values of mobilities at room temperature for various compound semiconductors are as follows:

GaAs electron mobility	$8000 \, cm^2/Vs$
GaAs hole mobility	$400 \, cm^2/Vs$
CdTe electron mobility	$1000 \, cm^2/Vs$
CdTe hole mobility	$70 \, cm^2/Vs$
$CuInSe_2$ electron mobility	$\sim 1000 \, cm^2/Vs$

These values are on the order of mobility values for elemental semiconductors. Thin film values are as follows:

GaAs epi-electron mobility	$2680 \, cm^2/Vs$
GaAs epi-hole mobility	$400 \, cm^2/Vs$
GaAs poly-hole mobility	$<10 \, cm^2/Vs$
CdTe epi-electron mobility	$100 \, cm^2/Vs$
CdTe epi-hole mobility	$50 \, cm^2/Vs$
CdTe(p) poly-electron mobility	$300 \, cm^2/Vs$
CdTe(p) poly-hole mobility	$5 \, cm^2/Vs$
$CuGaSe_2$ epi-electron mobility	60–$250 \, cm^2/Vs$ (values range according to degree of Cu excess concentration)
$CuGaSe_2$ poly-electron mobility	$20 \, cm^2/Vs$
$Cu(Ga,In)Se_2$ epi-hole mobility	200–$300 \, cm^2/Vs$
$Cu(Ga,In)Se_2$ poly-hole mobility	See Figure 1.5

These values also are comparable to the values for their counterparts in elemental semiconductors although the maximum values reported are for the most part smaller than those for the elemental semiconductors at equivalent charge carrier density,

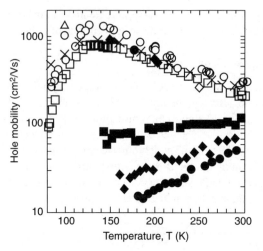

Figure 1.5. Hole mobility as a function of temperature for various epitaxial films of Cu(Ga,In)Se$_2$. Film plane orientations as follows: open circles {112}; X {002}; black filled points {220}. (From A. Rockett, Final Report, NREL/Sr-520-34335(2003) with permission.)

except those based on the GaAs system. Since this comparison holds when the main contribution to charge scattering is due to phonons then this result must be contributed to the greater phonon scattering[143] present in the compound semiconductors listed above. In part, this result, as well as economic factors account for the use of the elemental semiconductors in those transistor applications where mobility is a significant criterion. However, the maximum mobility occurs in the GaAsX compound semiconductors which accounts for its use despite its cost.

1.2.4. Effect of defects on mobility of amorphous semiconductors.

It is believed that the drift mobility in amorphous semiconductors is limited by the temporary trapping of charge carriers and can be described by the relation:

$$\mu_D = \mu_o/(1 + f_{trap}) \qquad (1.8)$$

where f_{trap} is the ratio of the time that the carrier spends in localized traps to that spent in mobile states or the ratio of the concentration of carriers in traps at any one time to the free carrier density (in extended states). For a single trap level at ΔE_T below E_C (the conduction band edge, for electrons) the parameter f_{trap} is given by:

$$f_{trap} = (N_T/N_C)\exp(\Delta E_T/kT) \qquad (1.9)$$

where N_C and N_T are the densities of conduction band states (at E_C) and trapping states (at E_T), respectively. For multiple trapping site levels, the relations become somewhat more complicated, but the qualitative result is the same in the sense that the higher is the trap density the lower is the drift mobility. Street,[28] based on various data, has estimated that μ_o for electrons is not larger than about 20 cm^2/Vs for hydrogenated amorphous silicon. In state-of-the-art a-Si:H electron mobilities approach 20–30 cm^2/Vs,[40] which suggests that f_{trap} is much less than unity in these films, i.e. the traps are already full.

1.3. Effect of structure on mobility in semiconductor superlattices.

Superlattices may be engineered or occur naturally. An elegant example of an engineered superlattice designed to produce high mobility uses one layer devoid of scattering centers, other than thermal ones, to contain and transport the electrons and an adjacent layer containing dopants to provide the free electrons. The materials are chosen so that the conduction band of the latter layer must empty its electrons into the conduction band of the high mobility layer.

An example of a naturally occurring superlattice is the CuAuI structure of InGaAs where the cation sites in alternate (001) planes are solely occupied by In in one plane and by Ga in the adjacent plane. The result of such long range ordering of the cations is to remove the small displacements of ions from their lattice sites that exist in the unordered structure due to the difference in atomic size between the two cations and, hence, the deleterious effect of such displacements on scattering of the electrons moving along (001) planes. Such effects on the mobility along (001) planes in InGaAs due to long range order have been measured.[91]

1.4. Effect of structure on conductivity and mobility in organic thin films.

The flexibility of organic thin films and low cost for large area application provide the basis for their use. Although their main application to date has been as organic light emitting diodes there is also a great interest in using them as flexible thin film transistors. Thus, charge carrier mobility in these films will be explored below.

Molecular and grain structure affect conductivity and mobility in amorphous and crystalline organic films, respectively. In particular, the conducting polymers consist of a backbone having alternating single and double bonds resulting in a "π-conjugated network". A necessary structural condition for high conductivity is that disordered regions separating ordered regions be short enough to prevent localization of the electrons and that ordered regions consisting of straight polymer strands be plentiful. The conductivity can be varied many orders of magnitude in these materials by doping.[92]

In thin films formed from oligomers, charge transport can be adversely affected by the lack of crystalline order[93] and in crystalline films by the presence of grain boundaries.[94a] It is speculated that the abnormally large hopping distance at the grain boundary is responsible for the greatly lowered mobility compared to that observed in single crystals. Values of about $1\,cm^2/Vs$ have been measured for the time-of-flight hole mobility in single crystals of aromatic hydrocarbons of the acene series.[94b] Films of such single crystals, if produced economically, may compete with a-Si:H films for TFT applications. Thus, the crystalline content or degree of order in conjugated polymers and oligomers represents the prime structural parameter which affects both the conductivity of metallic polymers and the mobility of semiconducting oligomers.

The highest mobility for the crystalline oligomeric acenes achieved to date, of the order of $1\,cm^2/Vs$, applies to room temperature values and not to low temperature values (see Figure 1.6). As shown, in good single crystals both hole and electron mobility approach values larger than $10\,cm^2/Vs$ as temperature decreases below 50 K. This result suggests that for monocrystalline films means of restricting the thermal motion

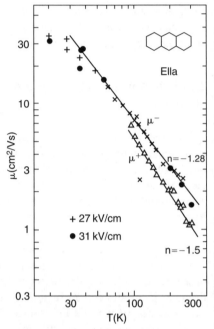

Figure 1.6. Electron and hole mobilities as a function of temperature for anthracene. (Reproduced from N. Karl et al., Synth. Met. 41–43, 2473–2481(1991). Copyright 1991 with permission from Elsevier.)

of the molecules that opens up the gap between neighboring molecules (by decreasing their free volume or by increasing the bonding between oligomer tips), without affecting their electronic states, might succeed in raising the room temperature mobility of these organic systems. In this regard, a single crystal of C_{60} yielded an electron mobility of $6000\,cm^2/Vs$,[126] which is not inconsistent with the concept in the previous sentence since the C_{60} molecules are close-packed in a fcc lattice.

1.5. Charge carrier densities.

Defect structure affects the conductivity of semiconductors not only via the mobility, but also by its effect on the charge carrier densities n and p in equation (1.1). In particular, in many compound semiconductors the antisite defect structure, as affected by deviation from stoichiometry, controls the charge carrier densities. Also, the presence of deep level defects can alter charge carrier densities substantially. Indeed, deep-level-trapping point defects are often introduced deliberately in order to produce insulating grade semiconductors.[107] Often their unintentional presence harmfully lowers charge carrier density.[101] Charge carrier densities can be reduced also by passivation of donors by compensating impurity type defects, such as hydrogen.[102] For example, at temperatures above about 600°C, soluble atomic hydrogen acts to reduce the carrier density in Mg doped GaN, but does not affect it in Si-doped GaN.[103]

Various schemes have been developed with the objective of increasing the dopant solubility in semiconductors. One scheme is based on replacing a random distribution of the dopant by one in which the dopants are restricted to parallel planes in the semiconductor which are separated from each other. This scheme is believed to result in a decrease in the rate of formation of dopant precipitates at the film interfaces.[104] Another scheme based on band theory uses the property of band edge displacement with thickness in type II superlattices to cause deep level traps to either act as donors, or to promote semi-insulating behavior, or to act as shallow acceptors.[105]

2A. Minority carrier diffusion length.

For many solar cell absorber films the minority carrier diffusion length (L) is an important parameter. It depends on the density of the effective non-radiative recombination centers. We will consider the relationship between these two parameters below. In an hydrogenated amorphous silicon p-type thin film positively charged dangling bonds that are homogeneously distributed are the main candidate for these defects although recombination also should occur at neutral dangling bonds. In a poly-Si p-type thin film again positively charged dangling bonds, but ones that are located at grain boundaries, stacking faults, twins and dislocations, are the defects believed to provide the limiting recombination centers for charge carriers. (In many cases impurities act to provide these centers, but we are focussing in this section primarily on intrinsic defects.) In the following sections we provide simplified derivations of minority carrier diffusion lengths for each of several types of thin film materials, not including monocrystalline layers. The latter do not differ from bulk monocrystalline materials and derivations of the corresponding DL can be found in many texts on the subject (e.g. A.H. Fahrenbruch and R.H. Bube, in **Fundamentals of Solar Cells**, Academic Press, New York, 1983).

2A.1. Hydrogenated amorphous silicon

We make the assumption that the major contribution to the recombination rate of carriers at room temperature arises from recombination at dangling bonds.

Recombination can occur both at charged dangling bonds and at neutral dangling bonds. In the absence of an electric field the diffusion length L can be shown to be equal to $(kT\mu_D\tau)^{1/2}$ using the relation $L = (D\tau_F)^{1/2}$, the Einstein relation $\mu_o = eD/kT$ and the relation $\mu_o\tau_F = \mu_D\tau$, where μ_o is the free carrier mobility, μ_D is the drift mobility, τ_F is the free carrier lifetime and τ is the total lifetime for deep trapping. Now it has been shown that for hydrogenated amorphous silicon the product $\mu_D\tau N_D$ is a constant, where N_D is the number of corresponding type dangling bonds.[41] This constant varies slightly for the different recombination possibilities at dangling bonds. According to Street[42] these products are $2.5 \cdot 10^8, 5 \cdot 10^7, 4 \cdot 10^7$, and $1.5 \cdot 10^7 \text{cm}^{-1}\text{V}^{-1}$ for the recombination of an electron at a neutral dangling bond, at a positively charged dangling bond, for a hole at a neutral dangling bond and at a negatively charged dangling bond, respectively. Using a typical density of about $3 \cdot 10^{15} \text{cm}^{-3}$ for each type of dangling bond we obtain corresponding diffusion lengths equal to $4.66 \cdot 10^{-5}, 2.08 \cdot 10^{-5}, 1.87 \cdot 10^{-5}$, and $1.14 \cdot 10^{-5} \text{cm}$.

These results may be compared to those of Sakata et al.[123] shown in Figure 1.7 for the dependence of diffusion length on the neutral dangling bond defect density in hydrogenated amorphous silicon. Two aspects of the analysis given above may be checked with these data. First is the inverse dependence of diffusion length on the square root of the defect density given by $L = (kT\mu_D\tau)^{1/2} = (kTC/N_D)^{1/2}$ that is obtained by substituting $\mu_D\tau = C/N_D$. The dashed line in Figure 1.7 shows the predicted dependence for the diffusion length of holes corresponding to recombination at neutral dangling bonds using Street's value for the corresponding constant C.

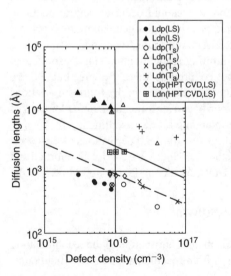

Figure 1.7. Diffusion length versus defect density in a variety of samples of a-Si:H. The points denoted in the upper right with subscripts n and p denote electron and hole diffusion lengths, respectively. The defect density was varied in a variety of ways including light soaking (LS) (solid symbols), changing the substrate temperature T_s (open symbols), and changing the annealing temperature T_a (crosses). The points corresponding to the HPT CVD designation (bottom two lines in the list of symbols) were obtained from samples deposited via a method different from that used for the samples corresponding to the other points. (Reprinted with permission from I. Sakata, M. Yamanaka and T. Sekigawa, J. Appl. Phys. 81, 1323(1997). Copyright 1997 American Institute of Physics.)

As shown the theoretical line is consistent with the corresponding data. The predicted dependence for electrons is shown by the full line. In this case the predicted line is below, but parallel to most of the corresponding data, except for the data represented by plus signs inserted in open squares. These two sets of data for electrons stem from two different modes of processing the a-Si:H films. The reason for the discrepancy between them is unknown. Thus, we may conclude that our analysis which yielded that the diffusion length depends inversely on the square root of the dangling bond density is confirmed by the data while the value of C corresponding to mobile electrons recombining at neutral dangling bonds given by the uppermost data in Figure 1.7 does not agree with Street's value, whereas the data corresponding to the other method of producing the a-Si:H films do correspond to a C value in agreement with Street's value.

The diffusion lengths in Figure 1.7 are significant to the use of a-Si:H as solar cell absorbing layers in that the thicknesses of such layers usually exceed these DL values. This result accounts for the fact that light absorption in such amorphous layers always occurs in the presence of an electric field because under such circumstance the drift length replaces the diffusion length and is given by:

$$L_{dr} = \mu E \tau = qL^2 E/kT \qquad (1.10)$$

where E is the electric field strength. For a reasonable electric field strength of 10^4 V/cm, L_{dr} equals 0.5 μm for the smallest L value obtained above. Thus, for $E \geq 10^4$ V/cm the drift length is sufficient to minimize recombination at dangling bonds in a-Si:H absorber layers 0.5 μm thick.

It may be of use later in this book to have values of the capture cross-sections corresponding to the various traps. These can be evaluated by noting that from random walk theory the diffusion coefficient D can be related to the trap density N_D and the free carrier lifetime τ_F by the relation $D = nr^2/2\tau_F = (\sigma N_D)^{-1} r/2\tau_F$, where n is the number of diffusive jumps in the time τ_F and r is the jump distance. (The volume swept out by a free carrier in its diffusive random walk is approximately given by $nr\sigma = (N_D)^{-1}$, where σ is the capture cross-section of the trap and N_D is the volume density of such traps.) But $D = kT\mu_o/e$. Hence, $\sigma = er/(2kT\mu_D\tau N_D)$. Using Street's values of the product $\mu_D\tau N_D$ given above, the values of the capture cross-section corresponding to recombination of an electron at a neutral dangling bond, a positive dangling bond, and of a hole at a neutral dangling bond and a negative dangling bond are $3.6 \cdot 10^{-15}$, $1.8 \cdot 10^{-14}$, $2.25 \cdot 10^{-14}$, and $6.0 \cdot 10^{-14}$ cm^2, respectively. (It should be mentioned that a later numerical analysis, based on dual-beam photoconductivity measurements of the steady-state photoconductivities and sub-band-gap absorption yield the respective values of 10^{-15}, 10^{-15}, 10^{-16}, and 10^{-15} cm^2 for these capture cross-sections.[43] Adoption of the latter would require use of the numerical model of these investigators, which would not serve our purpose of revealing clearly the effect of defects on electrical properties. Hence, we have chosen to present Street's method in this analysis.)

Although recombination at room temperature occurs mainly at dangling bonds, at lower temperatures recombination takes place via a variety of modes. Street[44] has given a detailed description of such recombination, which will not be described here because it does not bear upon our objective of relating properties to defect structure.

Illumination of a-Si:H generates additional dangling bond defects – the Staebler–Wronski effect – which act to degrade a-Si:H solar cells, in part by increasing the recombination center population and thereby decreasing the DL. In this context it is interesting to note that although a-Si$_x$Ge$_{1-x}$:H suffers from the Staebler–Wronski effect, microcrystalline (μc)-Si$_x$Ge$_{1-x}$:H alloys do not display a Staebler–Wronski effect, as manifested by a decay of the photoconductivity under intense illumination.[96] If this observation is reproducible, then it suggests that in this microcrystalline material the dangling bonds exist primarily at the grain boundaries and that light exposure cannot generate additional dangling bonds at the grain boundaries, i.e. both the crystalline grains and the grain boundaries are structurally stable with respect to perturbations induced by light exposure whereas the amorphous network is not stable in this regard.

It is interesting that the same controversy over the intrinsic or extrinsic nature of the dangling bond traps we encountered in the previous section also exists in the amorphous silicon community. Redfield and Bube,[26] for example believed that the dangling bond defects present in a-Si:H are each associated with carbon, oxygen, and/or nitrogen atoms, although the dangling bond itself belongs to a Si atom adjacent to the impurity atom. This viewpoint offers the hope that should some technique of refining the a-Si:H of these impurities be developed then the density of dangling bonds would be capable of being markedly decreased.[45] However, should the interpretation of the previous paragraph be correct then this hope is a fantasy. Ref. [144] cites the evidence against impurity traps.

2A.2. Effect of recombination centers at dislocations.

Deep-level traps located at dislocations can act as recombination centers in both monocrystalline and polycrystalline silicon. There are other sources of recombination centers in both single crystals and polycrystalline materials. However, at this point let us consider the contribution of dislocations to the diffusion length in effectively single crystal silicon. We will assume that there are n_t deep-level traps per unit length along a dislocation line on the average and N_d dislocations per unit area. We will further assume that these traps are homogeneously distributed throughout the volume so that we may make use of the relation for the diffusion length developed in the previous section in the form $L = (r/2\sigma N_D)^{1/2}$ where we have used the relation between $\mu_D \tau$ and σN_D also derived in the previous section. In the present application $N_D = 3n_t N_d + N_0$, where N_0 is the

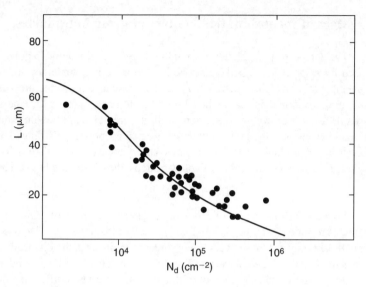

Figure 1.8. Experimental values of the effective diffusion length as a function of dislocation density. (Reproduced with permission from El Ghitani and Marinuzzi, J. Appl. Phys. 66, 1723(1989).) The solid line is a plot of equation (1.11).

density of recombination centers present in the material in the absence of dislocations. Thus,

$$L = [r/2\sigma(3n_tN_d + N_0)]^{1/2} \tag{1.11}$$

Now, El Ghitani and Martinuzzi[47a] have evaluated L (electron) values for large grained silicon as a function of dislocation density. Their data are shown in Figure 1.8. The line shown corresponds to the following values: $L_0 = [r/2\sigma N_0]^{1/2} = 70\,\mu m$ and $\sigma n_t = 9 \cdot 10^{-9}\,cm$. The first value is quite reasonable since it is known from the work of Pizzini et al.[47b] that below a dislocation density of $10^3\,cm^{-2}$ the effect of dislocations on the diffusion length is not measureable. The value of σN_0 deduced from L_0 is then $2.4 \cdot 10^{-4}\,cm^{-1}$. We will assume a value for the capture cross section corresponding to positive dangling bonds that we evaluated in the previous section of $1.8 \cdot 10^{-14}\,cm^2$. From this the value of n_t is then evaluated to be $5 \cdot 10^5\,cm^{-1}$. This value may be compared with $5 \cdot 10^5\,cm^{-1}$ obtained by Wilshaw and Fell[50] for dislocations in hexagonal loops on the {111} plane of silicon, the value of $2 \cdot 10^6\,cm^{-1}$ measured by Wilshaw et al.[48] from an analysis of EBIC contrast from screw dislocations as a function of beam current and temperature and that of $4 \cdot 10^6\,cm^{-1}$ measured by Omling et al.[49] We conclude that since there is at least one independent value of n_t equal to that deduced from the data of Figure 1.8 then our result for n_t is reasonable and the analysis on which it is based is likewise reasonable. The traps giving rise to the data in Figure 1.8 are primarily impurity atoms at dislocations.

2A.3. Effect of recombination centers at grain boundaries.

The defects responsible for most of the charge recombination in poly-Si thin films used for solar cells are believed to be located at grain boundaries, at least for films having grain sizes less than 1000 μm. These grain boundaries are between columnar grains oriented normal to the film plane. However, because the charge carrier in its diffusive step follows a random walk it is reasonable to assume that the recombination centers at the grain boundaries are homogeneously distributed. In this case we can simply follow the derivation that led to equation (1.11) with the exception that we substitute for N_D, the volume density of traps, the expression N_{gb}/d. Thus,

$$L = (rd/6\sigma N_{gb})^{1/2} \tag{1.12}$$

Figure 1.9 shows a plot of diffusion length as a function of grain size. It is important, as revealed in Figure 1.7 and in the work of Pizzini et al.,[51b] to use diffusion length data only for those cases where in large grain sizes the dislocation density is less than $10^4 cm^{-2}$. As shown, in Figure 1.9 the data obey the dependence of the diffusion length on the square root of the grain size with two different values of the capture cross-section. The value of σN_{gb} for the polycrystalline and multicrystalline data equals 320. Since N_{gb} cannot exceed the number of atom sites at grain boundaries per unit area, $\sim 10^{15} cm^{-2}$ then σ for these grain boundary recombination sites must exceed $3.2 \cdot 10^{-13} cm^2$. Similarly, the value of σ for the microcrystalline

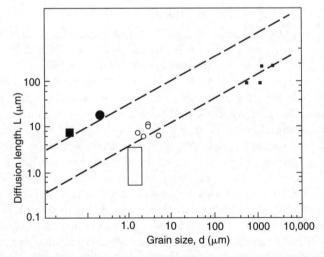

Figure 1.9. Diffusion length as a function of grain size. Open circles are from Ref. [52], small closed squares are from Ref. [51b]. The larger filled square point represents microcrystalline Si and the filled circle laser crystallized a-Si:H. The dashed lines have a slope equal to 0.5. See text for meaning of rectangular box.

and laser recrystallized a-Si:H grain boundaries must exceed $3.2 \cdot 10^{-15}\,\text{cm}^2$. Thus, we obtain the result that for Si the trapping site at the polycrystalline and multicrystalline grain boundaries has a capture cross-section on the order of that of a charged impurity in crystalline Si, whereas those at the microcrystalline and laser recrystallized grain boundaries have a cross-section on the order of dangling bonds in a-Si. Actually, the capture cross-section at the latter grain boundaries must be larger than that for a dangling bond because unless the density of recombination traps at these grain boundaries exceeds that in the bulk there would be no grain size effect for these boundaries on the diffusion length.

After the above was written, the writer discovered the work of Taretto et al.[127] who reported a similar plot except that instead of interpreting the different lines in terms of different capture cross-sections did so in terms of different recombination velocities at grain boundaries. Their plot is shown in Figure 1.10. The concept underlying both figures is the same and the independent results may be considered as support for it.[142]

2A.4. Polycrystalline compound semiconductors.

From remarks in the literature[128] one might think that grain boundaries in compound semiconductors are less prone to induce recombination than those in elemental semiconductors. However, I have not been able to find any evidence to substantiate this hypothesis. For example, it is known that measured diffusion lengths in good quality material vary between 0.5 and $3\,\mu\text{m}$ for CdTe[37] and CuInSe$_2$[54] for films that have a grain size between about 1 and $2\,\mu\text{m}$. A box that encompasses these data is shown in Figure 1.9. From the position of this box just the opposite hypothesis would be supported. This subject is discussed more fully in Chapter VII.

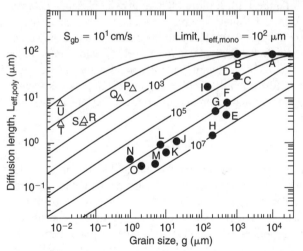

Figure 1.10. Diffusion length versus grain size for various recombination velocities. (Reprinted with permission from K. Taretto et al., J. Appl. Phys. <u>93</u>, 5453(2003). Copyright 2003 American Institute of Physics.)

2B. Solar efficiency and diffusion length.

2B.1. Silicon cells.

At the outset it is important that in analyzing solar efficiency data one must determine whether the diffusion length is longer or shorter than the absorber layer thickness. In the event that diffusion length is longer than the absorber layer thickness then one would expect that all the incident photons would reach the external circuit. In this event one would expect, since the incident radiation has a constant intensity for the same conditions of test, that the short-circuit current would be independent of the grain size and of the diffusion length. In fact, as shown in Figure 1.11, this is validated by the collected data. Hence, any systematic variation for this constraint on diffusion length/absorber layer thickness of the efficiency must be due to the open circuit voltage.

When the relation between diffusion length and absorber layer thickness is reversed we should expect that the short-circuit current would depend upon diffusion length and grain size. As shown in Figure 1.12 this expectation is obeyed by the data. (For both figures the data were produced by a variety of different sources and scatter in these results should thus be expected, as revealed. However, the conclusions drawn above are not dependent on this scatter.)

Now, we plot for the same cells as for Figure 1.12 the cell efficiency versus the diffusion length in Figure 1.13. A plot of the same cell efficiency data versus grain size shows much greater scatter. Now an explanation of the dependence of both the short-circuit current and cell efficiency on the diffusion length is in order.

For the case for which the thickness of the absorber layer is much larger than both the minority carrier diffusion length and the optical absorption length at the chosen wavelength, and for which the field strength is nil in the quasi-neutral p-type layer adjacent to a p–n homojunction, the short-circuit current at the depletion layer edge is given by:[31]

$$J_{SC} = q\Gamma/(1 + 1/\alpha L_n) \qquad (1.13)$$

where Γ is the photon flux, α the absorption coefficient, and L_n the minority carrier diffusion length. This relation assumes that the incident photons generate electron-hole pairs with a depth dependence of $\alpha e^{-\alpha z}$ and that the probability of carriers at the depth z reaching the depletion layer edge and thereby contributing to the short-circuit current is somewhat complicated but leads upon integration to equation (1.13). We now plot the inverse of equation (1.13) in Figure 1.14. From a least mean-square fit we determine the values of $q\Gamma$ and α to be $33\,mA/cm^2$ and $1.5 \cdot 10^4\,cm^{-1}$, respectively. The latter value is quite acceptable in view of the absorption coefficient for silicon shown in Figure 1.15 at the wave-length of the maximum intensity in the solar spectrum (0.5 microns). Thus, equation (1.13) yields

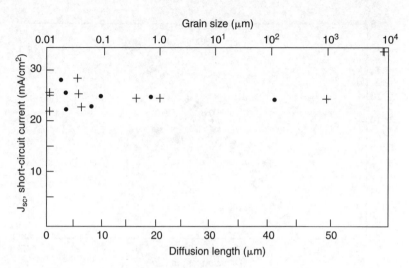

Figure 1.11. Showing independence of short-circuit current on grain size and diffusion length for solar cells in which diffusion length exceeds absorber layer thickness. Grain size, +; diffusion length, filled dots. (Data from Ref. [129].)

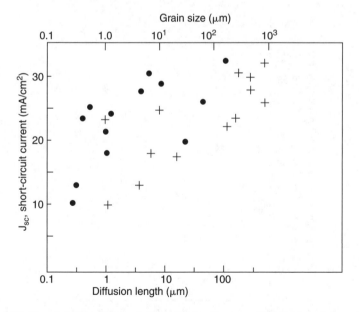

Figure 1.12. Showing dependence of short-circuit current on diffusion length for solar cells in which diffusion length is less than absorber layer thickness. Grain size, +; diffusion length, filled dots. (Data from Ref. [129].)

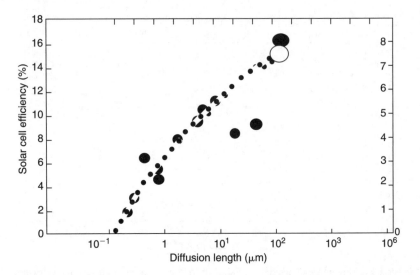

Figure 1.13. Solar cell efficiency for cells having diffusion length smaller than absorber layer thickness; experimental values represented by filled circular points (left ordinate); dotted line corresponds to equation (1.16) and right ordinate. (Data from Ref. [129].)

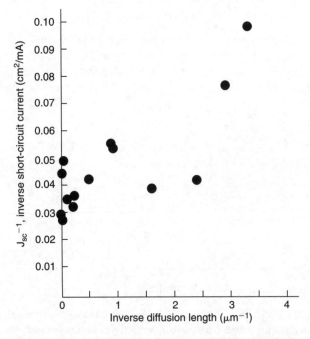

Figure 1.14. Inverse short-circuit current versus inverse diffusion length. (Data from Ref. [129].)

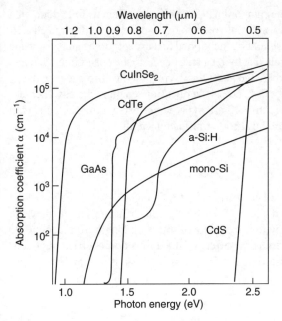

Figure 1.15. Absorption coefficient as a function of photon energy for various semiconductors. (Reprinted with permission from H.J. Möller, in **Semiconductors for Solar Cells**, Artech House, Inc., Norwood, MA, USA. http://www.artech-house.com)

a value of α and a dependence on the diffusion length that is consistent with experiment. Since the cell efficiency depends upon the product of the short-circuit current and the open-circuit voltage it is then reasonable for these solar cells to expect that the efficiency will also be dependent on the absorption coefficient and the diffusion length.

The open-circuit voltage of the solar cell is related to the short-circuit current by the relation:

$$V_{OC} = (kT/q)\ln[J_{SC}/J_o) + 1] \tag{1.14}$$

where for the above case

$$J_o = q(D_n/L_n)(n_i^2/N_A) \tag{1.15}$$

and D_n is the electron diffusivity in the p-type absorber layer, n_i is the intrinsic electron density in this layer, and N_A is the acceptor type dopant density for this layer. The solar cell efficiency is proportional to the $J_{SC} \cdot V_{OC}$ product. Thus, the solar cell efficiency, η:

$$\eta \sim [1 + (\alpha L_n)^{-1}]^{-1}\ln\{L_n/[1 + (\alpha L_n)^{-1}] + K\} \tag{1.16}$$

We have fitted this relation to the data at $L_n = 100\,\mu m$ to obtain a value for K (note that the value of $K = 5.31$ is based on using micron units for α^{-1} and L_n) and plotted the right-hand side of relation (1.16) as the dotted line in Figure 1.13 with the values

of this quantity indicated on the right ordinate in the figure. There is little doubt that
the dependence of this quantity on L_n mirrors that of the experimental solar cell efficiency for these solar cells. It should be mentioned that except for the smallest value
the dependence is almost entirely due to ln L_n and, hence, to the open-circuit voltage.
Thus, for poly-Si solar cells where the absorber layers limit the efficiency the dominant parameter controlling the efficiency is the diffusion length in the absorber layers. Points below the line in Figure 1.13 are likely to correspond to solar cells with
efficiency limited by other than the absorber diffusion length.

2B.2. Compound cells.

The measured values of cell efficiency for $CuInGaSe_2$ as a function of the
diffusion length are shown in Figure 1.16. For given diffusion length they are
higher than those for Si cells within the range of diffusion length from about 0.05
to 10 microns. We shall determine the effect of the absorption coefficient at constant diffusion length on the efficiency. We keep L constant in equation (1.16),

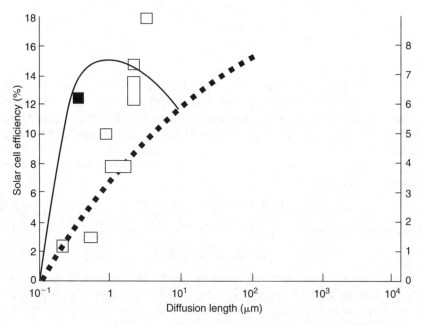

Figure 1.16. Points are experimental values for $CuInGaSe_2$ films. Full line represents
theoretical dependence of efficiency for $CuInGaSe_2$ films on diffusion length taking into
account absorption coefficient of $CuInGaSe_2$. Dotted line corresponds to dotted line in
Figure 1.13 and values for poly-Si. Points below full line may have efficiency limited by
other than diffusion length of absorber layer.

note that the values that enter the constant K are roughly unchanged between Si and the chalcopyrite cells, and form the ratio of efficiencies for the two types of cells. This ratio is given by:

$$\frac{[1 + (\alpha_2 L_n)^{-1}]\ln\{L_n/[1 + (\alpha_1 L_n)^{-1}] + K\}}{[1 + (\alpha_1 L_n)^{-1}]\ln\{L_n/[1 + (\alpha_2 L_n)^{-1}] + K\}}$$

Using this ratio, setting $\alpha_1 = 10^5\,\text{cm}^{-1}$, $\alpha_2 = 10^4\,\text{cm}^{-1}$, and the values of the efficiency along the dotted line for Si in Figure 1.13, the full line shown in Figure 1.16 was calculated and plotted. We may tentatively conclude that the higher values of efficiency for the chalcopyrite solar cells relative to the Si based cells in the range of diffusion length from about 0.1 to 10 microns is a consequence of the higher absorption coefficient of the former. (The position of the line in Figure 1.16 depends sensitively upon the value of the absorption coefficient for Si; a smaller value than that chosen will raise the maximum and shift it to the right.) Of course, there are many other factors that affect the efficiency of solar cells. However, their consideration is outside the scope of this book which concentrates upon the effect of structure on properties. Even the effect of absorption coefficient is not an effect of structure. Perhaps the main conclusion one may draw from the above analysis is that the diffusion length significantly affects the solar cell efficiency. Structural defects limit the diffusion length. For some reason the diffusion length in the chalcopyrite absorber layers has not at this writing exceeded about 3 μm. We will discuss these defects in more detail in Chapter VII.

3. Leakage current.

Leakage currents in devices can occur either in the active regions of devices in the semiconductor lattice itself or through the thin insulators separating regions of different electric potential. One of the sources of leakage current in the active regions of devices is the presence of extended defects in such regions. In insulators leakage current is usually related to the presence of point defects rather than extended defects. We will first discuss the properties of extended defects that may lead to enhanced conductivity due to their presence in the active regions of semiconductor devices and then we will discuss how various point defects act to enhance the leakage current in insulators.

3.1. Leakage current due to extended defects in semiconductors.

Extended defects in semiconductors have the potential to become paths having higher conductivity than exists along paths through the bulk material for several reasons. In germanium it has been found that quasi-metallic conduction occurs along large angle grain boundaries of the $\Sigma 11$-structure in n as well as p-type with a carrier density of from 5 to 6 · 10^{12} cm^{-3}. However, no excess conductivity was detected along a Σ-25-type (<100> tilt axis, 16.3° tilt angle) grain boundary in n-type silicon having a dopant concentration of 10^{13} cm^{-3}.[59]

One dimensional conduction along one type of dislocation (unidentified) has been found in Ge with the conductance given by 6.9 · 10^{-11} exp(-0.022/kT) Ω^{-1}m, where kT is in units of eV.[59] Microwave conductivity in both n- and p-type silicon was found to be enhanced parallel to dislocation lines that were introduced by plastic deformation at low temperature (800°C), as long as carrier trapping at other defects was negligible, i.e. the density of defects was less than the dopant density.[60] The enhanced conductivity was very weakly temperature dependent. The authors explained their results in terms of one shallow acceptor and one shallow donor dislocation band that becomes partly filled and conducting in the dark only when the Fermi level is fixed near the band edges. Dislocations have been found to enhance the leakage current in memory cells in silicon thereby decreasing the time required between successive refresh cycles. The temperature dependence of this relation yielded an activation energy of about 0.5 eV.[58]

The leakage current in MOS power devices correlates to the density of oxidation induced stacking faults in the presence of metal impurities, as shown in Figure 1.17. Mahajan[61] has described several examples from the literature where dislocations have been responsible for leakage currents leading to failure of devices. In particular, Cd doped InP photodiodes[62] suffered severe shorting of the junction when the dislocation density in the film exceeded about 2.2 · 10^7 cm^{-2}. In this case no linear correlation between dislocation density and reverse current at a given reverse voltage was found. Rather, the reverse current

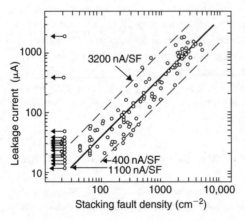

Figure 1.17. Leakage current at 400 V reverse bias versus the stacking fault density for MOS-power devices. The arrows at points to the left of the figure indicate that the stacking fault densities are below the detection limit. For most of the data the leakage current is proportional to the stacking fault density. (Reproduced with permission from Kolbensen et al., in **Structure and Properties of Dislocations in Semiconductors**, eds. S.G. Roberts et al., IOP Bristol, 1989. Copyright 1989 IOP Publishing Ltd.)

exhibited a two-fold order of magnitude increase with less than a one order of magnitude increase in dislocation density above about a dislocation density of $2.2 \cdot 10^7 \, cm^{-2}$, whereas below this density a threefold order of magnitude change in the dislocation density produced less than one order of magnitude change in the reverse current.

As another example, p–n diodes formed at various positions cutting across a swirl pattern in silicon revealed reverse currents ranging from 10 to $10^5 \, nA$ at a reverse voltage of 25 V. Swirl patterns have been shown to consist of two types of defects, one of which is a prismatic dislocation loop.

From the brief summary of the properties of extended defects given above we can deduce several possible effects of these defects. One is that they may provide additional charge carriers due to the segregated metal impurities that can ionize and act as donors or acceptors with less activation energy required in the vicinity of the extended defects. Secondly, in some cases there is evidence that the mobility along the extended defects in semiconductors may be higher than in the bulk material, i.e. the conduction may be quasi-metallic along the extended defects. Third, there is the possibility of diffusion of dopants along the dislocation lines threading p- and n-type regions causing a p-type pipe to thread through an n-type region or vice versa. Thus, there is sufficient justification for steps to prevent the presence of extended defects in active regions where they can contribute to the leakage current or to add to currents that must be minimized.

The reader who is alert may have made a connection between the essential relationship of this section, and after having read the previous subsection, with solar efficiency. Leakage current in solar cell absorber layers is a shunting current that acts to reduce the solar efficiency below that corresponding to the diffusion length of the absorber layer. The effect of structure induced shunts in the absorber layer is at this writing just being recognized. The use of two independent measurements – light beam induced current (LBIC) to make visible the position of defects such as grain boundaries and infrared thermography to make visible the position of regions of high shunting currents – have shown that inversion layers along grain boundaries crossing the bulk of the absorber layer are the dominant structural shunting regions.[130] Further, investigation using X-ray microprobe fluorescence scanning revealed metal impurities segregated along such shunting regions.[131] The similarity between these results and those described above for other devices is obvious. Thus, it is to be hoped that solar cell and device researchers concerned with leakage current problems can aid each other to solve their problems.

With the miniaturization of devices it may be possible to make use of the enhanced conductivity of extended defects in semiconductors. In particular, it may be possible to use a dislocation line itself as an active conductor region in a FET arrangement where the current along the dislocation is modified by a gate between source and drain connections to the dislocation line. Contrary to the examples described above this one would be a beneficial leakage current.

3.2. Leakage current in gate oxides and other insulators.

Leakage currents are also of interest in gate oxides which are quite thin, such as those used in EEPROMS. In such a device, write/erase cycles involve the passage of current through the gate oxides. The passage of such current pulses induce defects in the oxide film that trap charge and that act to increase low level pre-tunneling leakage currents. Indeed, the latter parameter was found to be proportional to the volume density of the charge traps.[65] One component of the low level leakage current was determined to arise from the discharging of these traps and the other component was found to be due to Schottky barrier emission. The current component from the latter increases with the density of traps in the oxide. The accumulation of such write/erase cycle induced traps in the oxide leads to wear out of the oxide and failure of the device. These leakage currents provide a limiting effect on the scaling down of EEPROM tunnel oxide thicknesses.[65] Incidentally, the trap densities providing these measureable leakage currents were on the order of from 10^{18} to 10^{19} cm^{-3}. Similar trap densities have been reported in silicon oxynitrides. For example, a dangling bond density of $2 \cdot 10^{19}$ cm^{-3} has been reported in hydrogenated silicon oxynitrides containing less than 0.7 O/(O + N), but it should be noted that these dangling bonds are situated at a variety of species.[66] However, wearout, as measured by the increase in leakage current for a given voltage stress cycle, was found to be less in silicon oxynitride gates than in pure silicon oxide gates despite the fact that the oxynitride gates had higher trap densities at the beginning of stressing, i.e. although the increment in the increase in leakage current per increment in trap density was the same for both materials the silicon oxide had a higher leakage current at a given trap density and the increment in trap density per voltage stress cycle was greater for the oxide than for the oxynitride. These results are illustrated in Figure 1.18. The higher increment in leakage current and trap density per voltage cycle for the oxide than for the oxynitride is not related to the initial trap density, which is lower in the oxide. The origin of this effect is not known.

In ultra-thin gate SiO_2 oxides (<35 Å) there is no charge build-up in traps because the charge easily tunnels out of the traps and the limiting mode on the leakage current becomes interface-trap-assisted tunneling.[132] Two mechanisms have been identified in ultra-thin gate oxides: field ionization tunneling at low field strengths and Fowler–Nordheim tunneling at high field strengths.[134] These leakage currents are the cause of what is known as "disturbs" in floating gate non-volatile memories.

Higher dielectric constant oxides, such as HfO_2, are being considered as replacement for SiO_2 gate oxides having an effective thickness less than 1 nm at this writing. HfO_2 is a crystalline oxide. One effect of polycrystallinity on properties is to produce a non-homogeneous character to the electrical properties of the gate oxide as may be produced by differential reaction between the gate oxide and the silicon of differently oriented grains. It has been estimated that this effect can bring about a reduction of charge carrier mobility in the transistor of 35%.[137] Replacement of the crystalline oxide by an amorphous $HfAlO_X$ oxide was found

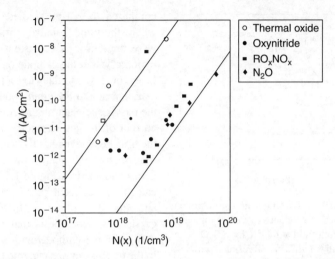

Figure 1.18. Increase in leakage current density versus trap density. Note that although the increase in leakage current density per increment in trap density is the same for oxide as for oxynitride materials, the increment in trap density per voltage stress cycle was higher for the oxide than for the oxynitrides. (From J.T. Richardson et al., MRS Symp. Proc. 284, 357(1993) with permission.)

to reduce the leakage current by two orders of magnitude. The HfO_2 leakage current is orders of magnitude smaller than for SiO_2 at the same thickness. At this writing no mechanism studies regarding leakage current in the higher dielectric constant oxides has appeared to the best knowledge of the author.*

3.3. Leakage current in ferroelectric films.

Leakage is also a failure mechanism operating in ferroelectric films, such as $BaSrTiO_3$, used as capacitors. It occurs by the electromigration of positively charged oxygen vacancies toward the cathode. In the early stages of leakage it is believed[76a] that the leakage current is determined by tunneling through the grain boundary barriers which presumably offer the greatest resistance to the migration of charged oxygen vacancies at the start of application of an electric field. In time the interface between cathode and oxide becomes the limiting resistance. As the oxygen vacancy concentration builds up at the oxide/electrode interface the resistance due to this interface decreases and allows the leakage current to increase at increasing rates to the onset of failure. The observation that a capacitor with a columnar grain structure had an order of magnitude higher leakage current as

---

* See comments with respect to HfO_2 interfaces in Chapter VI.

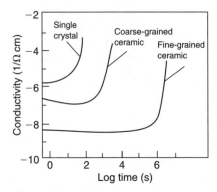

Figure 1.19. (a) Effect of microstructure on dependence of conductivity versus time at a 0.625 kV/cm field at 600°K for SrTiO$_3$ doped with 0.1 at.% Ni. (Reprinted with permission from R. Waser, in **Ferroelectric Ceramics**, eds. N. Setter and E.L. Colla, Birkhauser Verlag AG, Basel. Copyright 1993 Birkhauser Verlag AG.)

compared to a fine-grained structure where the grain boundary orientations were random[63a] is consistent with the hypothesis that the cathode/oxide interface provides an initially lower resistance than that due to the grain boundaries to the passage of electromigration induced current of oxygen vacancies. However, the time dependence of the leakage current prior to the sharp upward regime needs explanation. Figure 1.19(a) indicates three different behaviors dependent upon the grain size. In the single crystal the behavior is that described above for the cathode/oxide interface. In the coarse-grained ceramic the resistance initially decreases while in the fine-grained sample the initial resistance remains constant for some lengthy period. However, the same three types of behavior can be induced with increasing dc electric field in a system consisting of BaTiO$_3$–0.5%MgO–x%HoO$_{3/2}$ as shown in Figure 1.19(b) with samples having the same grain size. In the opinion of this writer the last word concerning the mechanism(s) controlling the leakage current behavior of these ferroelectric ceramics is yet to be written.

A bilayer film consisting of a polycrystalline layer and an amorphous one showed a lower leakage current than the polycrystalline one alone.[63b] After some

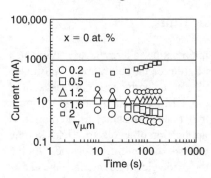

Figure 1.19. (b) Leakage current versus time for BaTiO$_3$–0.5%MgO doped with Ho. (From Key Eng. Mater. <u>248</u>, 183(2003) with permission.)

time the leakage current in the latter begins to rise rapidly, as shown in Figure 1.19(a), leading to voltage breakdown, a topic considered in the next section. This phenomenon of rapid current rise is called "resistance degradation" in the literature and, as noted previously, is believed to be a consequence of the drift of doubly charged oxygen vacancies to in front of the cathode with a concomitant depletion of this defect in front of the anode. Due to the coupled equilibria determining the local defect chemistry at every point of the dielectric there is an increase in the hole concentration in front of the anode and an increase in the electron concentration in front of the cathode. This process makes the capacitor behave like a forward biased p–n junction, thus accounting for the enhanced current density.[76c] The use of IrO_2 as a bottom electrode for a $Pb(Zr,Ti)O_3$ (PZT) ferroelectric capacitor decreased the leakage current with respect to that obtained with a Pt bottom electrode.[76d] It was believed that the improvement is due to the absorption of the oxygen vacancies by the metallic oxide electrode. However, the bottom electrode affects the texture and grain size of the PZT deposited on it,[76e] both of which affect the leakage current. Nevertheless, the beneficial effect of oxide electrodes in combatting resistance degradation appears incontrovertible.[111]

Dopants that act to decrease the concentration of oxygen vacancies are believed to exert a beneficial effect in inhibiting resistance degradation. Implantation of Mn into barium strontium titanate acted to lower the leakage current by a factor of 10, but the mechanism by which the Mn exerts this effect is not known.[113]

The conductivity–microstructure relationship of $Bi_4Ti_3O_{12}$ is different from that above for the Ba titanates. In particular, epitaxially grown $Bi_4Ti_3O_{12}$ has a lower conductivity ($0.28 \cdot 10^{-8}\Omega^{-1}$) than a film having a granular microstructure ($2 \cdot 10^{-8}\Omega^{-1}$), but with a preferred (104) orientation, at low electric fields. Also, the leakage current for the granular film is higher than that for the epitaxial film.[76f] Also, $SrBi_2Ta_2O_9$ layered perovskites consisting of two $SrTaO_3$ planes alternating with a Bi_2O_2 plane and related layered structures resist resistance degradation as is manifested by the absence of n-type depletion layers at the interfaces with the electrodes.[64]

Summarizing, the leakage current at the moment of resistance degradation in the perovskite thin films depends upon the perovskite/electrode interface and, in particular, upon the local structure (micro- and defect) at the interfaces between the ferroelectric layer and the electrodes. Resistance degradation in thin films involves the migration of defects in bulk to the perovskite/electrode interfaces. The leakage current prior to resistance degradation may be affected by grain size.

4. Breakdown voltage in thin film insulators.

The process of voltage breakdown in SiO_2 has been studied extensively and is fairly well understood at this time. It is due to hot electron ($>2\,eV$ above the bottom of the oxide conduction band) generation of interface defects that are trapping sites and is defect mediated.[68] At a given field strength across the insulator there is a critical interface charge above which breakdown occurs. Some of this charge has its origin in defects already present before voltage stressing. Thus, proper processing may enhance the resistance of a SiO_2 film to voltage breakdown. In particular it has been suggested that reduction of the number of oxygen vacancies near the interfaces,

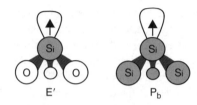

Figure 1.20. The E′ and the P_b defects at the Si/SiO$_2$ interface.

which it is believed can be done by intentional oxygen annealing or via nitridation schemes using ammonia (NH$_3$),[69] will accomplish this objective. Also, minimizing hydrogen incorporation and hydrogen motion in the oxide layer through the blocking action of oxynitride barriers created by some of these nitridation schemes would also reduce trap creation effects and increase device lifetime, particularly at high fields.[70]

If possible, lowering the operating temperatures of devices below 200°K should also reduce defect generation and increase charge to breakdown.[71]

What are the nature of the defects that act to trap charge and lower breakdown voltage? The dangling bond is the primary defect. But, there are different dangling bond defects. One is the E′ center illustrated in Figure 1.20, which consists of unpaired electrons in sp^3-hybridized orbitals of Si atoms backbonded to three oxygen atoms, and which is a hole trapping site. The E′ center is typically associated with a positively charged oxygen vacancy. Another is the P_b center, also illustrated in Figure 1.20, which consists of an unpaired electron on a Si atom bonded to three other Si atoms at the silicon/dielectric interface. This defect is responsible for the majority of interface states in as-processed Si/SiO$_2$ interfaces. Also, there are water species related electron traps that correlate to the hydrogen contents of nitrided oxides. Hydrogen is believed to play a Janus-like role, passivating dangling bonds and depassivating Si—H bonds to reform dangling bonds.

The mechanism by which the hot electrons produce defects is apparently twofold. First, both charge species, holes and electrons, are involved, albeit in different regimes. In one case trap creation at the cathode interface occurs when hot electrons with energy greater than 2 eV release a mobile species, such as hydrogen, near the anode interface which can move to and interact within the cathode interfacial region to produce defects that then act to trap electrons.[68] Another process of interface defect generation occurs additionally in relatively thick oxides (>15 nm) that involves impact ionization by the hot electrons through the bulk of the oxide producing holes that then move to near the cathode interface where they generate defects by the recombination reaction between free electrons and trapped holes. The assumption that breakdown occurs when a sufficient number of defect states is generated at the cathode interface yields a dependence of interface-charge-to-breakdown on average electric field in the oxide that mirrors the experimental relationship.[72]

In very thin films another phenomenon known as quasi-breakdown occurs that is characterized by a sudden increase at low field in leakage current and noise. This breakdown occurs when the interface trap density reaches a constant value which increases with film thickness.[135] At a gate oxide thickness of 37 Å this critical value is ~0.8–1.5 · 10^{11} cm^{-2}. For a thickness of 45 Å this value is 8 · 10^{11} cm^{-2}.

Silicon oxynitrides have higher breakdown voltages than SiO$_2$.[73] They also have higher initial populations of dangling bond species but lower rates of defect generation per given voltage stressing. However, the dangling bond defect density in the oxide upon voltage stressing soon exceeds that in the oxynitride upon continued cycles of voltage stressing.[67] This result is consistent with the association between a critical number of interface-defect states and voltage breakdown.

The mode of dielectric breakdown for the higher dielectric constant gate oxides, such as HfO$_2$, has not been studied sufficiently at this writing, but

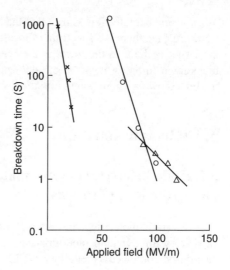

Figure 1.21. Breakdown time versus field in lead zirconium titanate (x), BST, ($\triangle$), and strontium titanate (o). (From J.F. Scott et al., MRS Bull. 21, 33(1996) with permission.)

one study suggests that the limiting factor is breakdown of the intermediate layer between the gate oxide and silicon.[139]

Of the dielectric materials currently being considered for DRAM capacitor application, BaTiO$_3$ thin films were reported to have breakdown fields between 0.5 and 1 MV/cm,[74] as compared to 4–10 MV/cm for SiO$_2$, 8 MV/cm for SiN, and 8 MV/cm for SiO$_2$/SiN stacked films.[73] The breakdown field in BaTiO$_3$ is determined by the onset of an exponential current dependence on electric field. Conduction in BaTiO$_3$ obeys a relation of the form:

$$J = A \exp[-(Q - E)/kT]$$

where E is the applied field strength. Given a fixed charge q to breakdown with t the time to produce this charge during which the field is applied, then through $J = q/t$ and the exponential dependence of J on the field strength we obtain the relationship shown in Figure 1.21, from which the breakdown field strength for (BaSr)TiO$_3$ is deduced to be about 0.5 MV/cm, i.e. charge will leak through the capacitor in less than 1 h. It is believed that the roughness of the electrode/insulator interface may effect the onset of breakdown. A mechanism for this breakdown hypothesizes the growth of dendrites of shorted material starting at the interfaces between electrodes and insulator.[136] Thus, the mechanism of the effective breakdown strength in the perovskite type materials differs from that of the silicon based dielectrics.

Incidentally, it is known that the breakdown field depends upon the electrode area, i.e. there is a spectrum of localized breakdown field strengths such that the larger is the area the lower is the breakdown field for the film.[76b] Since the breakdown initiates at the electrode/insulator interface the chance of initiating the breakdown at more potent initiation sites increases with electrode area.

5. Dielectric constant.

5.1. High dielectric constant materials in ferroelectric and piezoelectric applications.

High dielectric constant materials also have application as thin film capacitors for dynamic random-access memories (DRAM), non-volatile ferroelectric random-access memories (NVFRAM), piezoelectric films for MEMS, for tunable microwave components, phase shifters, resonators, etc. Ferroelectric materials having the perovskite structure are being considered for these applications. This structure is illustrated in Figure 1.22 and derives its high polarization from the fact that the structure does not have inversion symmetry. For the specific case of $Pb(Zr, Ti)O_3$ shown in Figure 1.22 this lack of inversion symmetry corresponds to the off body-centered-cubic equilibrium position of the Zr,Ti ions. There are six such equilibrium points for these ions in the unit cell and the point occupied determines the direction of polarization and the c-axis of the tetragonal unit cell.

In order to function as a memory device the capacitor unit must be able to be switched from one polarization to the reverse polarization. Hence, the capacitor

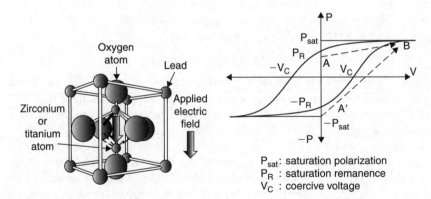

Figure 1.22. Perovskite unit cell of the ferroelectric PZT and a typical polarization hysteresis loop that characterizes the ferroelectric capacitor.

unit in the device must have one orientation of the polarization vector at the start and be capable of being switched to the reverse polarization. Considering that the capacitor film is constrained to a substrate, that the film itself is subject to stress induced by both the transformation strain on being cooled through the Curie temperature separating one crystal structure (e.g. cubic) from another (e.g. tetragonal) and by the strain due to the difference in thermal expansion coefficient between film and substrate, it is extraordinary to expect the film to have one polarization in its minimum free energy state and not to switch back spontaneously to this minimum free energy state after it has been switched away from it. Indeed, the latter process gives rise to the phenomenon known as "retention loss". These ferroelectric films may be set into an average single state of polarization by poling, i.e. applying a sufficiently high electric field at an elevated temperature to switch all the domains to an average of one polarization. The random grain orientations prevent a strictly single polarization vector but do allow a narrow spread about one close to the normal to the film plane.

The main structural parameter acting for ferroelectric and piezoelectric films is the texture.[138] The remanent polarization and the e_{33} strain are maximized for a film that approaches a single crystal texture.[128] Indeed, for a polar film of AlN the properties improve as the half-width of the (0001) rocking curve X-ray diffraction peak decreases[129] while for an epitaxial ferroelectric film having a (001) texture the remanent polarization equals that of a bulk single crystal.[128] However, the coercive field strength needed to switch the polarization is much higher for a ferroelectric film[128] than for an isolated island about 1 μm on a side and 1 μm high.[140] The latter result is a consequence of the clamping effect of the substrate for the film and its relative absence for the isolated island.

Another factor that is important for properties in these films is the state of internal stress. The latter is affected by many variables, such as difference in thermal expansion coefficients between substrate and film, and allied variables. Tensile internal stress tends to rotate the hysteresis loop clockwise leading to low remanent polarization and piezoelectric response.

Grain size is a structural parameter that affects the properties of these films. It has been found that grains smaller than 1 micron do not exhibit 90° domains.[141] Hence, these films do not have an extrinsic contribution to their remanent polarization or piezoelectric response. The larger the grain size the larger is the remanent polarization and piezoelectric response.

Switching occurs by domain wall motion. This motion can be pinned by defects (oxygen vacancies) and by grain boundaries.[115] Domain wall pinning is responsible for the phenomenon of "ferroelectric fatigue", which is manifested as a narrowing of the hysteresis loop with switching cycles. That domain wall motion occurs has been substantiated directly by piezoelectric force microscopy.[140] The fact that use of metallic oxide electrodes eliminates ferroelectric fatigue implies that the doubly charged oxygen vacancies developed in perovskite ferroelectrics using metal electrodes and not in those using metallic oxide electrodes are the domain wall pinning centers in the former switching capacitors.

The clamping effects due to the substrate and polycrystallinity are minimized by controlling the related film parameters. In particular, increase in film thickness decreases the effect of substrate clamping. Also, increasing the grain size similarly reduces the clamping due to misorientations of domains across grain boundaries – indeed, at the limit in grain size achieved in an epitaxial film the maximum possible remanent polarization is achieved. Both substrate and grain misorientation clamping is minimized also in small ferroelectric islands roughly 1 micron cubed in volume.

Summarizing, the effect of structure upon the properties of ferroelectric capacitor and piezoelectric films is manifested as an effect of texture, of grain size, of defects at interfaces, and of substrate constraints.

5.2. Low dielectric constant materials.

The need to decrease the delay time (dependent upon the resistance–capacitance product, RC) along interconnections has led not only to the use of lower resistivity interconnection material (copper) but also to a search for low dielectric constant materials suitable for interlevel and interline dielectric insulation in multilevel integration.[110] The latter direction is necessary because the line to line capacitance increases with decrease in the feature size.[111]

The standard material that has been used until recently is SiO_2 with a dielectric constant in the range between 3.8 and 4.2. Polymeric thin films have lower dielectric constants ranging from 1.8 to 3.0. Still lower dielectric constants may be obtained with porous materials. These data provide the hint as to one possible effect of structure on the dielectric constant. Covalently bonded material is composed of polarizable species and space not containing electrons. Porous material obviously contains electron-free space, which has unity dielectric constant. Further, as the number of atoms in unit volume of a homogeneous covalently bonded material increases the dielectric constant will also increase as one may deduce from the Lorenz–Lorentz equation. Indeed these concepts account for the main trend in dielectric constant between SiO_2, polymeric films and porous materials listed above.

However, dielectric constant is but one of the many criteria needed to be satisfied by conducting line insulating material. A sophisticated analysis of the problems associated with the choice and/or design of a low dielectric constant line insulator is provided in Ref. [144]. At this writing the effect of structure on the needed properties has not been researched although there have been a few studies relating to molecular structure and its effect of dielectric properties. For example, Boese et al.[77a] have found that when the precursor, poly(amic) acid dissolved in N-methylpyrrolidone is spun onto a surface and then imidized thermally to form poly(p-phenylene biphenyltetracarboximide), denoted as BPDA-PDA, the dielectric constant of the resulting film is higher in the film plane (3.5) than normal to it (2.8). The polymer chains in this film were aligned preferentially in the film plane by the spinning process. This anisotropy in dielectric constant is characteristic of many polyimides.[77b,c] These polymer chains have pi electron clouds along the chain axis contributing a larger polarizability parallel to the chain than normal to it.

Labadie et al.[78] found it possible to reduce this anisotropy in the dielectric constant by preparing a different polyimide derived from 2,2'-bis(trifluoromethyl)benzidine, denoted as BTFB. This polyimide, PMDA-BTFB, exhibited a much reduced anisotropy of the dielectric constant, an in-plane value of 2.9 versus an out-of-plane value of 2.6. Labadie et al. chose BTFB based on the fact that the backbone of this polymer has a pronounced twist ($\approx 90°$) at the biphenyl linkage, thus assuring some out-of-plane polymer chains. Anisotropy in the dielectric constant is undesirable because it can lead to crosstalk noise between signal lines in the same wiring plane.

Despite the advantage of lower dielectric constant, the current insulator on copper lines is not an organic but is a modified SiO_2. This has been a consequence of the inadequate mechanical strengths associated with the organic based dielectric and its interfaces, which have been insufficient to prevent delamination in the course of processing. An analysis of the requirements for resistance to delamination[79] concludes that it relates to the stresses developed during chemical–mechanical polishing of the stacks of lines, insulators, etc. and that the main factor resisting delamination is the area per insulator that is not laterally surrounded by stiffer material. The smaller is this area the smaller the stress applied to the insulator and its interfaces and the higher is the resistance to delamination. Of course this is a design consideration and not a material one. The strength of the insulator and its interfaces comprises the material-related aspect. An analysis of interface strength[80] suggests the obvious, that it increases with increasing bond strength across the interface and with increasing number of bonds across unit area of the interface.

6. Effects of structure on the critical current density of superconducting thin films.

High T_C superconducting thin films represent an intricate tapestry of relationships between structure and electrical properties. With respect to critical current density, this class of superconducting thin films is characterized by a very short coherence length, smaller than the thickness of most grain boundaries. This is made apparent by the dependence of critical current density on the tilt angle between adjoining grains about the tilt axis common to both grains[82] and by separate measurements of the critical current density within the regions of the grain and grain boundary.[83] Figure 1.23 shows the former dependence for a tilt grain boundary with close to a [001] tilt axis in $YBa_2Cu_3O_{7-x}$ where the tilt axis is normal to the film plane. (The [001] directions of the adjoining grains were not precisely parallel having a misorientation of φ in the plane normal to the common grain boundary and less than 1° in the plane parallel to the grain boundary.) Table 1.2 lists the critical

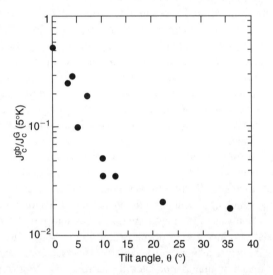

Figure 1.23. Ratio of the across grain boundary critical current density to the average value of the critical current density in the adjoining grains versus the tilt angle with a [001] tilt axis, as measured at 4.2–5°K. (Modified with permission from D. Dimos, P. Chaudhari, J. Mannhart and F.K. LeGoues, Phys. Rev. Lett. 61, 219(1988). Copyright 1988 by the American Physical Society.)

Table 1.2. Critical current density (10^3 A/cm^2) at 4.2–5°K

θ (φ)		Grain 1	Grain 2	Grain boundary
0°	(3°)	7140	8000	4000
3°	(7°)	5900	5300	1400
4°	(4°)	270	220	73
5°	(9°)	6000	5700	560
7°	(4°)	140	180	40
10°	(4°)	7800	8000	410
10°	(4°)	7000	6100	240
12.5°	(2°)	3800	3400	160
22°	(2°)	800	260	11
35.5°	(4°)	1350	1400	25

current densities within each grain and across the grain boundary as a function of the tilt angle θ and tilt axis misorientation φ. Now it is well known that tilt grain boundaries consist of edge dislocations parallel to the tilt axis with the distance between dislocations d related to the tilt angle θ by the relation $\theta = d/b$, where b is the Burger's vector of the dislocation. The tilt boundary consists of regions of good fit separated by regions of poor fit. It is believed that the regions of good fit carry the superconducting current while the regions of poor fit are normal resistivity regions. This interpretation is in agreement with recent measurements[84] showing a significant critical current density in a magnetic field of several Tesla. It has been found that subjecting such a tilt boundary to ozone increases the superconducting critical current density.[85] This fact has been interpreted to signify an oxygen depletion at the grain boundary of the as-deposited film. Where along the boundary this depletion occurs is not known. Although there have been studies involving attempts to introduce impurity atoms selectively along grain boundaries in such films the results of these experiments are not edifying in terms of defining the sites along the boundary which the impurities occupy.[85]

Another effect of structure in the high T_C superconducting films is related to the anisotropy of the coherence lengths. The coherence length perpendicular to the CuO_2 planes in these materials is much smaller than parallel to these planes. Thus, the orientation of the films is an important parameter. The art of controlling the orientation has been extensively developed.[86]

Defects other than grain boundaries can affect the superconducting properties. In the high T_C films the Cu–O layers are the paths for the paired charge carriers. Hence, oxygen deficiencies degrade the critical current density, especially at grain boundaries. Also, cation disorder has been found to decrease T_C of YBCO films.[89] However, cation disorder may be responsible for the pinning of magnetic flux in these films.[88]

Since integration of high T_C films with Si is important to the potential applications of such films much attention has been paid to methods to achieve such integration. All these methods involve use of an oxide layer between film and the Si to provide a substrate of the high T_C film. The surface character of the substrate can affect the structure of the superconducting film and the resulting properties.[90] The art and science of producing these integrated substrates and buffer layers is undergoing rapid development at this writing.

The primary applications of thin film high T_C superconductors is in the area of electronics. Among these potential applications are interconnects, Josephson junction devices such as sensors, and transistors. Even at substantial magnetic fields, the critical currents at 77°K are in excess of $10^6 A/cm^2$ in thin films due to some unknown defect in the film.[88] Thus, applications in magnetic fields for these thin films are possible. Also, with recent development of techniques of depositing highly oriented YBCO thin films on thin oxide buffer layers, which themselves are on stainless steel tape the potential of obtaining high critical current density flexible tape for power cable and magnet applications is much closer to attainment. Critical current densities of $10^6 \, A/cm^2$ have been obtained in such tapes and the dependence of critical current density on tilt angle shown in Figure 1.23 has been reproduced in these composite tapes.[89]

Recapitulation

The grain boundary is the main defect acting to produce electrical resistivity values that are higher in as-deposited metallic thin films than in bulk materials. This occurs because the grain size in as-deposited metallic thin films is quite small, on the order of several tens of nm. In monocrystalline semiconductors, the majority charge carrier mobility in thin films is decreased by thermal scattering, scattering by ionized impurities and by charged dislocations. Charged dislocations exert a greater scattering effect on carriers moving perpendicular to the dislocation lines than parallel to them and are more effective scatterers than the equivalent number of ionized impurity atoms. Grain boundaries, by trapping majority charge carriers, act as barriers to the transport of majority charge in semiconductors. The dependence of the majority charge carrier mobility on grain size can be used to distinguish between the effects of intragranular defects and grain boundaries on this mobility. The majority charge carrier mobility in amorphous hydrogenated silicon is limited by the temporary trapping of the majority charge carriers.

The minority charge carrier diffusion length in semiconductors varies as the square root of the grain size and inversely as the square root of the dislocation density in accord with the predictions of a model for minority charge recombination at these defects. Because the diffusion length in a-Si:H is smaller than the absorption length which, in turn, is smaller than the drift length, solar cells of this material are designed to absorb light in regions of non-zero electric field. The effect of diffusion length on open-circuit voltage accounts for the dependence of solar efficiency on diffusion length for polycrystalline silicon for cells for which the diffusion length is larger than the absorber layer thickness and the absorption length. The solar efficiency increases at constant diffusion length with decreasing absorption length. Intracrystalline defects are believed, at present, to limit the diffusion length and efficiency of compound semiconductor based solar cells.

Leakage current in insulating grade semiconductors is produced by the presence of extended defects. In insulators, leakage current is affected by the density of traps that can release charge under voltage at the operating temperature, but the quantitative effect of a given trap density on leakage current depends upon the particular insulator material. These traps are likely to be dangling bonds at Si atoms and are generated by the passage of current through the insulator. Current leakage in polycrystalline ferroelectric perovskite thin films is limited by interfaces. Resistance degradation in these films is accelerated by the presence of metal electrodes at which depletion layers can form by the pile up of oxygen vacancies and is eliminated with the use of metallic oxide electrodes.

Breakdown voltage in SiO_2 thinner than 15 nm is deleteriously affected by the presence of hydrogen in the film and in thicker films by the interaction of hot electron generated holes with the Si/SiO_2 interface. In both cases dangling bonds are generated at the cathode interface which act to trap electrons there. Breakdown occurs upon the development of a critical charge density at the cathode interface. Voltage breakdown in ferroelectric perovskite films depends upon the effect of electric field on leakage current.

The dielectric constant and remanent polarization of ferroelectric perovskite films depend upon the main domain orientation of the film, which is affected by the film stress but primarily by the texture. Mobile oxygen vacancies and grain boundaries act to pin domain wall motion and thus help to bring about the disappearance of a hysteresis loop in the phenomenon of ferroelectric fatigue generated by continuous switching of film polarization. Clamping due to substrate constraint decreases with increase in thickness and grain size as well as by the removal of lateral constraints such as misoriented grains or the formation of isolated ferroelectric islands.

Finally, critical current density decreases markedly with increasing misorientation of grains in the polycrystalline high-temperature superconductor films.

References

1. (a) H.J. Wollenberger, in **Physical Metallurgy**, 3rd edition, eds. R.W. Cahn and P. Haasen, North-Holland, Amsterdam, 1983, p. 1146; (b) J.R. Lloyd and S. Nakahara, J. Appl. Phys. 48, 5092(1978); (c) From C.E. Botez et al., MRS Symp. Proc. 749, W8.1.1(2003).
2. (a) S.M. Rossnagel and J.J. Cuomo, Thin Solid Film. 171, 143(1989); (b) D.L. Dexter, Phys. Rev. 86, 770(1952).
3. J.W.C. de Vries, Thin Solid Film. 167, 25(1988).
4. J.W.C. de Vries, Thin Solid Film. 150, 201, 209(1987); J. Phys. F 17, 1945(1987).
5. J.R. Sambles, K.C. Elson and D.J. Jarvis, Phil. Trans. Roy. Soc. Lond. Ser. A 304, 365(1982); J.R. Sambles, Thin Solid Film. 106, 321(1983).
6. P. Ziemann and E. Kay, J. Vac. Sci. Tech. A1, 512(1983).
7. A.F. Mayadas, M. Shatzkes and J.F. Janak, Appl. Phys. Lett. 14, 345(1969).
8. E. Conwell and V.F. Weisskopf, Phys. Rev. 77, 388(1950).
9. W. Shockley, in **Electrons and Holes in Semiconductors**, D. Van Nostrand, New York, 1950.
10. M.Y. Ghannam, in **Advanced Silicon and Semiconducting Silicon-Alloy Based Materials and Devices**, eds. J.F.A. Nijs et al., IOP Publishing, Bristol, 1994, p. 413.
11. D.L. Dexter and F. Seitz, Phys. Rev. 86, 964(1952).

12. (a) R. Labusch and W. Schröter, in **Dislocations in Solids**, ed. F.N.R. Nabarro, North-Holland Publishing Co., Amsterdam, 1980, p. 127; (b) H.F. Matare, in **Defect Electronics in Semiconductors**, Wiley(Interscience), New York, 1971.

13. W.E. Taylor, N.H. Odell and H.Y. Fan, Phys. Rev. 88, 867(1952); M.L. Tarng, J. Appl. Phys. 49, 4069(1978); G.J. Korsch and R.S. Muller, Solid State Electron, 21, 1045(1978); R.K. Mueller, J. Appl. Phys. 32, 635, 640(1961); G.E. Pike and C.H. Seager, J. Appl. Phys. 50, 3414(1979); C.H. Seager and G. Cashner, J. Appl. Phys. 49, 3879(1978); G. Baccarani, B. Ricco and G. Spadini, J. Appl. Phys. 49, 5565(1978).

14. J.Y. Seto, J. Appl. Phys. 46, 5247(1975).

15. M.W.M. Graef, J. Bloem, L.J. Giling, F.R. Monkowski and J.W.C. Maes, **Proceeding of EC Photovoltaic Solar Energy Conference, 2nd**, Berlin, D. Reidel, Dordrecht, Netherlands, 1979, p. 65.

16. E. Campo, J.J. Pedroviejo, E. Scheid, D. Bielle-Daspet, A.Y. Messaoud, G. Sarrabayrouse and A. Martinez, MRS Symp. Proc. 303, 389(1993).

17. F. Okumura, K. Sera, H. Tanabe, K. Yuda and H. Okumura, MRS Symp. Proc. 377, 877(1995).

18. L.L. Kazmerski, W.B. Berry and C.W. Allen, J. Appl. Phys. 43, 3515(1972).

19. T. Baba, T. Matsuyama, T. Sawada, T. Takahama, K. Wakisaka and S. Tsuda, MRS Symp. Proc. 358, 895(1995).

20. T. Sameshima, M. Sekiya, M. Hara, N. Sano and A. Kohno, MRS Symp. Proc. 358, 927(1995).

21. C. Schwebel, F. Meyer, G. Gautherin and C. Pellet, J. Vac. Sci. Tech. B4, 1153(1986).

22. T. Serikawa, S. Shirai, A. Okamoto and S. Suyama, IEEE Trans. Electr. Dev. 36, 1929(1989).

23. R. Buchner, K. Haberger and B. Hu, in **Polycrystalline Semiconductors, Grain Boundaries and Interfaces**, eds. H.J. Möller, H.P. Strunk and J.H. Werner, Springer Proceeding in Physics 35, Springer-Verlag, Berlin, 1989, p. 289.

24. P.M. Lanahan, W.K. Schubert, Phys. Rev. B30, 1544(1984); N.M. Johnson, D.K. Biegelsen and M.D. Meyer, Appl. Phys. Lett. 40, 882(1982).

25. T. Tiedje, B. Abeles and J.M. Cebulka, Solid State Commun. 47, 493(1983); M. Günes, H. Liu, C.M. Fortmann and C.R. Wronski, IEEE 1st WCPEC, 1994, p. 512.

26. D. Redfield and R. Bube, MRS Symp. Proc. 325, 335(1994).

27. H.C. Card and E. Yang, IEEE Trans. Electr. Dev. ED-29, 397(1977); A.K. Ghosh, C. Fishman and T. Feng, J. Appl. Phys. 51, 446(1980).

28. R. Street, in **Hydrogenated Amorphous Silicon**, Cambridge University Press, Cambridge, UK, 1991, p. 311.

29. L.L. Kazmerski, in **Polycrystalline Semiconductors, Grain Boundaries and Interfaces**, eds. H.J. Möller, H.P. Strunk and J.H. Werner, Springer-Verlag, Berlin, 1989, p. 96.

30. A.P. Sutton, in **Structure and Properties of Dislocations in Semiconductors**, eds. S.G. Roberts, D.B. Holt and P.R. Wilshaw, IOP Conference Series 104, IOP, Bristol, 1989, p. 13.

31. A.L. Fahrenbruch and R.H. Bube, in **Fundamentals of Solar Cells**, Academic Press, New York, 1983, p. 364.

32. M. Kohyama, S. Kase and R. Yamamoto, MRS Symp. Proc. 262, 567(1992).

33. L.L. Kazmerski and Y.J. Juang, J. Vac. Sci. Tech. 14, 769(1977).

34. H.J. Möller, in **Semiconductors for Solar Cells**, Artech House, Norwood, MA, 1993, p. 292.
35. ibid., p. 286.
36. M.M. Al-Jassem, F.S. Hasoon, K.M. Jones, B.M. Keyes, R.J. Matson and H.R. Moutinko, **23rd IEEE Photovoltaic Specialists Conference**, 1993, 459.
37. A.D. Compaan, C.N. Tabory, Y. Li, Z. Feng and A. Fischer, ibid., 394.
38. J.R. Tuttle, M. Contreras, A. Tennant, D. Albin and R. Noufi, ibid., 415.
39. R. Noufi, R. Axton, C. Herrington and S.K. Deb, Appl. Phys. Lett. 45, 668(1984).
40. M. Günes, H. Liu, C.M. Fortmann and C.R. Wronski, IEEE 1st WCPEC, 1994, p. 512.
41. Ref. [31], p. 57.
42. R.A. Street, Phil. Mag. B49, L15(1984).
43. M. Günes, C.R. Wronski and T.J. McMahon, J. Appl. Phys. 76, 2260(1994).
44. Ref. [28], p. 278.
45. It was reported at the 1996 MRS Spring meeting that S. Yamasaki et al., have made electron paramagnetic resonance measurements that show that carbon and oxygen are not the cause of the Staebler–Wronski effect. See p. 70 of MRS Bull. 21, (1996).
46. A.K. Ghosh, C. Fishman and T. Feng, J. Appl. Phys. 51, 446(1980).
47. (a) H. El Ghitani and S. Martinuzzi, J. Appl. Phys. 66, 1723(1989).
48. P.R. Wilshaw, T.S. Fell and G.R. Booker, in **Point and Extended Defects in Semiconductors**, eds. G. Benedek, A. Cavallini and W. Schröter, NATO ASI Series B, Physics, 202, Plenum Press, New York, 1988, p. 251.
49. P. Omling, E.R. Weber, L. Montelius, H. Alexander and J. Michel, Phys. Rev. B32, 6571(1985).
50. P.R. Wilshaw and T.S. Fell, in **Structure and Properties of Dislocations in Semi-conductors**, eds. S.G. Roberts, D.B. Holt and P.R. Wilshaw, IOP Conference Series 104, IOP, Bristol, 1989, p. 85.
51. (a) S. Pizzini, F. Borsani, A. Sandrinelli and D. Narducci, in **Point and Extended Defects in Semiconductors**, eds. G. Benedek, A. Cavallini and W. Schröter, NATO ASI Series B, Physics, 202, Plenum Press, New York, 1988, p. 105; (b) S. Pizzini, A. Sandrinelli, M. Beghi, D. Narducci, F. Allegretti, S. Torchio, G. Fabbri, G.P. Ottaviani, F. Demartin and A. Fusi, J. Electrochem. Soc. Solid State Sci. Technol. 135, 155 (1988).
52. C. Feldman, N.A. Blum, H.K. Charles Jr. and F.G. Satkiewicz, J. Electron. Mater. 7, 309(1978).
53. Y. Hamakawa, IEEE 1st WCPEC, 1994, p. 34.
54. S. Mora and N. Romeo, J. Appl. Phys. 48, 4826(1977).
55. C. Ferekides, J. Britt, Y. Ma and L. Killian, in **23rd IEEE Photovoltaic Specialists Conference**, 1993, p. 389.
56. P.V. Meyers, T. Zhou, R.C. Powell and N. Reiter, in **23rd IEEE Photovoltaic Specialists Conference**, 1993, p. 400.
57. W.S. Chen, J.M. Stewart, W.E. Devaney, R.A. Mickelsen and B.J. Stanley, in **23rd IEEE Photovoltaic Specialists Conference**, 1993, p. 422.
58. B.O. Kolbesen, W. Bergholz, H. Cerva, F. Gelsdorf, H. Wendt and G. Zoth, in **Structure and Properties of Dislocations in Semiconductors**, eds. S.G. Roberts, D.B. Holt and P.R. Wilshaw, IOP Conference Series 104, IOP, Bristol, 1989, p. 421.

59. R. Labusch and J. Hess, in **Point and Extended Defects in Semiconductors**, eds. G. Benedek, A. Cavallini and W. Schröter, NATO ASI Series B, Physics, 202, Plenum Press, New York, 1988, p. 15.

60. M. Brohl and H. Alexander, in **Structure and Properties of Dislocations in Semiconductors**, eds. S.G. Roberts, D.B. Holt and P.R. Wilshaw, IOP Conference Series 104, IOP, Bristol, 1989, p. 163.

61. S. Mahajan, in **Semiconducting Materials and Related Technologies**, eds. S. Mahajan and L.C. Kimerling, Pergamon Press, Oxford, 1992, p. 85.

62. E.A. Beam III, H. Temkin and S. Mahajan, Semicond. Sci. Tech. 7, A229(1992).

63. (a) T. Makita, T. Horikawa, H. Kuroki, M. Kataoka, J. Tanimura, N. Mikami, K. Sato and M. Nunashita, MRS Symp. Proc. 284, 529(1993); J.F. Scott, B.M. Melnick, L.D. McMillan and C.A. Paz de Araujo, Integr. Ferroelectr. 3, 129(1993); V. Joshi, D. Roy and M.L. Mecartney, Integr. Ferroelectr. 6, 321(1995); C.J. Peng, H. Hu and S.B. Krupanidhi, Appl. Phys. Lett. 63, 1038(1993); (b) L.H. Chang, Q.X. Jia and W.A. Anderson, MRS Symp. Proc. 318, 501(1994).

64. J.F. Scott, F.M. Ross, C.A. Paz de Araujo, M.C. Scott and M. Huffman, MRS Bull. 21(7), 33(1996).

65. D.J. Dumin, J.R. Maddux and D.-P. Wong, MRS Symp. Proc. 284, 319(1993).

66. S. Viscaïno, Y. Cros and B. Reif, MRS Symp. Proc. 284, 339(1993).

67. J.T. Richardson, D.J. Dumin, G.Q. Lo, D.L. Kwong, B.J. Gross and C.G. Sodini, MRS Symp. Proc. 284, 357(1993).

68. D.J. DiMaria, E. Cartier and D. Arnold, MRS Symp. Proc. 284, 219(1993).

69. D.J. DiMaria, in **Insulating Films on Semiconductors**, eds. W. Eccleston and M. Uren, Adam Hilger, New York, 1991, pp. 65–72.

70. D.J. DiMaria and J.H. Stathis, J. Appl. Phys. 70, 1500(1991).

71. G.Sh. Gildenblat, C.-L. Huang and S.A. Grot, J. Appl. Phys. 64, 2150(1988).

72. D. Arnold, E. Cartier and D.J. DiMaria, Phys. Rev. B45, 1477(1992).

73. S. Scaglione, L. Mariucci, A. Mattacchini, A. Pecora and G. Fortunato, MRS Symp. Proc. 284, 345(1993); D. Waechter, T. Billard and P. Gogna, p. 443.

74. Q.X. Jia, J. Yi, Q. Shi, K.K. Ho, L.H. Chang and W.A. Anderson, MRS Symp. Proc. 284, 525(1993).

75. J.F. Scott et al., MRS Bull. 21, 33(1996).

76. (a) H. Chazono and H. Kishi, Jpn. J. Appl. Phys. 40, 5624(2001); (b) S. Matsubara, T. Sakuma, S. Yamamichi, H. Yamaguchi and Y. Miyasaka, MRS Symp. Proc. 200, 243(1990); ibid., 243, 281(1992); (c) R. Waser, T. Baiatu and K.-H. Hardtl, J. Am. Ceram. Soc. 73, 1645(1990); (d) M. Shimizu and T. Shiosaki, MRS Symp. Proc. 361, 295(1995); (e) I. Chung, J.K. Kee, W.I. Lee, C.W. Chung, S.B. Desu, MRS Symp. Proc. 361, 249(1995); (f) W. Jo, K.H. Kim, T.W. Noh, S.D. Kwon, B.D. Choe and B.D. You, MRS Symp. Proc. 361, 33(1995).

77. (a) D. Boese, S. Herminghaus, D.Y. Yoon, J.D. Swalen and J.F. Rabolt, MRS Symp. Proc. 227, 379(1991); (b) S. Herminghaus, D. Boese, D.Y. Yoon and B.A. Smith, Appl. Phys. Lett. 59, 104(1991); (c) F.S. Ip and C. Ting, MRS Symp. Proc. 381, 135(1995).

78. J.W. Labadie, H. Lee, D. Boese, D. Yoon, W. Volksen, P. Brock, Y. Cheng, M. Ree and K.R. Chen, **Proceeding of 43rd IEEE Electronic Components and Technology Conference**, 1993, p. 327.

79. V. McGahay, MRS Symp. Proc. 766, E6.1.1(2003).

80. M. Lane and R. Rosenberg, MRS Symp. Proc. 766, E9.1(2003).
81. (a) H. Tabata, H. Tanaka and T. Kawai, MRS Symp. Proc. 361, 453(1995); (b) B.A. Tuttle, T.J. Garion, J.A. Voight, T.J. Headley, D. Dimos and M.O. Eatough, in **Science and Technology of Electroceramic Thin Films**, eds. O. Auciello and R. Waser, Kluwer Academic Publishers, Dordrecht, Netherlands, 1995, p. 117; G.A.C.M. Spierings, G.J.M. Dormans, W.G.J. Moors, M.J.E. Ulenaers and P.K. Larsen, **Proceeding of 9th IEEE International Symposium on Applied Ferroelectrics**, 1994, 29.
82. D. Dimos, P. Chaudhari, J. Mannhart and F.K. LeGoues, Phys. Rev. Lett. 61, 219(1988).
83. P. Chaudhari, J. Mannhart, D. Dimos, C.C. Tsuei, J. Chi, M.M. Oprysko and Mm. Scheuermann, Phys. Rev. Lett. 60, 1653(1988).
84. E. Sarnelli, P. Chaudhari and J. Lacey, Appl. Phys. Lett. 62, 777(1993); M. Daümling, E. Sarnelli, P. Chaudhari, A. Gupta and J. Lacey, Appl. Phys. Lett. 61, 1335(1992).
85. J. Alarco, Yu. Boikov, G. Brorsson, T. Claeson, G. Daalmans, J. Edstam, Z. Ivanov, V.K. Kaplunenko, P.-A. Nilsson, E. Olsson, H.K. Olsson, J. Ramos, E. Stepantsov, A. Tzalenchuk, D. Winkler and Y.-M. Zhang, in **Materials and Crystallographic Aspects of HT$_C$ – Superconductivity**, ed. E. Kaldis, NATO ASI Series E 263, Kluwer Academic Publishers, Dordrecht, Netherlands, 1994, p. 471.
86. J. Mannhart, J.G. Bednorz, A. Catana, Ch. Gerber and D.G. Schlom, ibid., p. 453.
87. S.D. Lester, F.A. Ponce, M.G. Crawford and D.A. Steigerwald, Appl. Phys. Lett. 66, 1249(1995).
88. T.H. Geballe, MRS Symp. Proc. 341, 3(1994).
89. J. Ye and K. Nakamura, MRS Symp. Proc. 401, 429(1996).
90. P.R. Fletcher, C. Leach, F. Wellhofer and P. Woodall, MRS Symp. Proc. 341, 221(1994).
91. O. Ueda, MRS Symp. Proc. 417, 31(1996).
92. J. Tsukamoto, Adv. Phys. 41, 509(1992).
93. B. Servet, G. Horowitz, S. Ries, O. Lagorse, P. Alnot, A. Yassar, F. Deloffre, P. Srivastava, R. Hajlaoui, P. Lang and F. Garnier, Chem. Mater. 6, 1809(1994).
94. (a) C.D. Dimitrakopoulos, A.R. Brown and A. Pomp, J. Appl. Phys. 80, 2501(1996); (b) W. Warta and N. Karl, Phys. Rev. B32, 1172(1985); N. Karl, J. Marktanner, R. Stehle and W. Warta, Synth. Met. 41–43, 2473(1991); M. Pope and C.E. Swenberg, in **Electronic Process in Organic Crystals**, Oxford University Press, New York, 1982, p. 338; K. Bitterling and F. Willig, Phys. Rev. B35, 7973(1987).
95. A. Knierim, R. Auer, J. Geerk, G. Linker, O. Meyer, H. Reiner and R. Schneider, Appl. Phys. Lett. 70, 661(1997).
96. S.M. Cho, D. Wolfe, K. Christensen, G. Lucovsky and D.M. Maher, MRS Symp. Proc. 403, 471(1996).
97. R. Venkatasubramanian, B. O'Quinn, J.S. Hills, M.L. Timmons, D.P. Malta, J.B. Keyes and R. Ahrenkiel, MRS Symp. Proc. 403, 483(1996); C.M. Yang and H.A. Atwater, ibid., p. 113.
98. S. Goncalves-Conto, E. Müller, K. Schmidt and H. von Känel, MRS Symp. Proc. 402, 493(1996).
99. S.L. Zhang, J. Cardenas, F.M. d'Heurle, B.G. Svensson and C.S. Petersson, Appl. Phys. Lett. 66, 58(1995).
100. (a) See applicable papers in MRS Symp. Proc. 402 (1996); (b) C. Cabral Jr., L.A. Clevenger, J.M.E. Harper, F.M. d'Heurle, R.A. Roy, C. Lavoie and K.L. Saenger, Appl. Phys. Lett. (1977).

101. B. Chatterjee, S.A. Ringel, R. Sieg, I. Weinberg and R. Hoffman, MRS Symp. Proc. 325, 125(1994).

102. S.J. Pearton, C.R. Abernathy, W.S. Hobson and F. Ren, ibid., p. 261.

103. M.S. Brandt, N.M. Johnson, R.J. Molnar, R. Singh and T.D. Moustakas, ibid., p. 341.

104. J.E. Cunningham, W.T. Tsang, ibid., p. 247.

105. B. Monemar, P.O. Holtz, J.P. Bergman, Q.X. Zhao, C.I. Harris, A.C. Ferreira, M. Sundaram, J.L. Merz and A.C. Gossard, ibid., p. 3.

106. B.-S. Yoo, S.-G. Kim and E.-H. Lee, ibid., p. 279.

107. ibid., pp. 101, 261, 305 and 361.

108. P. Hiesinger, T. Schweizer, K. Kohler, P. Ganser, W. Rothemund and W. Jantz, J. Appl. Phys. 72, 2941(1992).

109. G.E. Stillman, S.A. Stockman, C.M. Colomb, A.W. Hanson and M.T. Fresina, MRS Symp. Proc. 325, 197(1994).

110. W.W. Lee and P.S. Ho, MRS Bull. 22, 22(1997).

111. N. Fukushima, K. Abe, M. Izuha and T. Kawakubo, MRS Symp. Proc. 493 (1997).

112. M.O. Aboelfotoh, H.M. Tawancy and L. Krusin-Elbaum, Appl. Phys. Lett. 63, 1622(1993).

113. J.D. Baniecki, R.B. Laibowitz, T.M. Shaw, P.R. Duncombe, D.A. Neumayer, M. Copel, D.E. Kotecki, H. Shen and Q.Y. Ma, MRS Symp. Proc. 493 (1998).

114. C. Basceri, S.E. Lash, C.B. Parker, S.K. Streiffer, A.I. Kingon, M. Grossman, S. Hoffman, M. Schumacher, R. Waser, S. Bilodeau, R. Carl, P.C. van Buskirk and S.R. Summerfelt, MRS Symp. Proc. 493 (1998).

115. A. Gruverman, S.A. Prakash, S. Aggarwal, R. Ramesh and H. Tokumoto, MRS Symp. Proc. 493 (1998); V. Gopalan and R. Raj, ibid.

116. A.L. Roytburd, MRS Symp. Proc. 474 (1998); K.S. Lee, Y.M. Kang and S. Baik, MRS Symp. Proc. 493 (1998).

117. (a) A.I. Kingon and S.K. Streiffer, MRS Symp. Proc. 493 (1998); (b) W.-J. Lee, H.-G. Kim and S.-G. Yoon, J. Appl. Phys. 80, 5891(1996).

118. B.-S. Yoo, M.A. McKee, S.-G. Kim and E.-H. Lee, Solid State Commun. 88, 447(1993).

119. T.-B. Ng, J. Han, R.M. Biefeld, J.C. Zolper, M.H. Crawford and D.M. Follstaedt, MRS Symp. Proc. 482 (1998) and many other papers in this symposium.

120. It has been suggested in a recent review paper by Birkmire and Eser in Annu. Rev. Mater. Sci. 27, 630(1997) that dopants, such as oxygen, that adsorb at grain boundaries in p-type CuInSe$_2$ and CdTe act to make the grain boundaries ineffective as recombination centers.

121. B. Cunningham, H. Strunk and D.G. Ast, Appl. Phys. Lett. 40, 237(1982).

122. K. Ismail, J. Vac. Sci. Tech. b14, 2776(1996).

123. I. Sakata, M. Yamanaka and T. Sekigawa, J. Appl. Phys. 81, 1323(1997).

124. A. Ourmazd, P.R. Wilshaw and G.R. Booker, Physica B116, 600(1983).

125. "As-deposited" signifies in this context deposition onto substrates which are at room temperature. When deposition is onto substrates held at elevated temperature then this fact will be noted in the book.

126. H. Sitter et al., Synth. Met. 138, 9(2003).

127. K. Taretto et al., Solid State Phenom. 93, 399(2003).

128. H. Morioka et al., MRS Symp. Proc. 784, C6.2.1(2004); S. Smith et al., Appl. Phys. Lett. 85, 3854(2004).

129. F.J. Hickernell, R.X. Yue and F.S. Hickernell, IEEE Trans. UFFC 44, 615(1997); H.P. Loebl et al., J. Electroceram. 12, 109(2004).
130. A.A. Istratov et al., NREL/CD-520-33586, 228(2003).
131. A. Kaminski et al., J. Phys. Condens. Matter 16, S9(2004).
132. W.Y. Loh, B.J. Cho and M.F. Li, Appl. Phys. Lett. 81, 379(2002).
133. www.veeco.com/appnotes/AN79_ElecChar_RevA0.pdf.
134. J.E. Park, J. Shields and D.K. Schroder, Solid State Electr. 47, 855(2003).
135. H. Guan et al., IEEE Trans. Electr. Dev. 48, 1010(2001).
136. H.M. Dinker and P.D. Beale, Phys. Rev. B41, 990(1990); H.M. Dinker, P.D. Beale and J.F. Scott, J. Appl. Phys. 68, 5783(1990).
137. G. Bersuker et al., MRS Symp. Proc. 811, D2.6.1(2004).
138. S. Trolier-McKinstry and P. Muralt, J. Electroceram. 12, 7(2004).
139. K. Torii et al., MRS Symp. Proc. 811, D2.7.1(2004).
140. V. Magarajan et al., Nat. Mater. 2, 43(2003).
141. B.G. Demczyk and R.S. Rai, J. Am. Ceram. Soc. 373 (1990); G. King and E.K. Goo, ibid., 6, 73(1990); M.O. Eatough et al., The Rigaku J. 12, 10(1995).
142. As noted in the discussion in Figure 1.9 it is the product of the capture cross-section and trap density that is deduced from the data in the figure. Taretto et al. interpreted the different recombination velocities to imply different trap densities. We have interpreted the shift in position of the lines in Figure 1.9 to imply different capture cross-section and apply this interpretation solely to the data in this figure because the processing of the microcrystalline films may well yield amorphous regions along grain boundaries.
143. S. Siebentritt and S. Schuler, J. Phys. Chem. Solid. 64, 1621(2003); L. Essaleh, S.M. Wasim and J. Galibert, J. Appl. Phys. 90, 3993(2001).
144. H. Fritzsche, Ann. Rev. Mater. Res. 31, 47(2001).

Problems

1. Why is the resistivity of a metal, monocrystalline film unaffected by a dislocation density of $10^8/cm^2$, whereas the same dislocation density in a monocrystalline semiconductor markedly increases the resistivity?

2. What defect is responsible for the excess resistivity in as-deposited metal films? What processing procedure will reduce the population of this defect?

3. What aspect of structure affects the resistivity of films thinner than the electron mean free path?

4. Comment on the effect of recombination centers on mobility of charge carriers?

5. Will surface scattering contribute to the electrical resistivity should electrons be specularly reflected from the surface?

6. For a dopant concentration of $10^{16}/cm^3$ and a grain size of 100 nm, use Seto's model to evaluate the expected electron mobility in poly-Si.

7. If you used a deposition procedure for poly-Si that yielded intragranular mobility equivalent to that of bulk monocrystalline silicon, which regime, that of equation (1.6) or that of equation (1.7), would you use to produce a poly-Si with the highest mobility?

8. What type of defects would you add to a semiconductor to produce an insulating region in the semiconductor? What type of defects would transform an insulating region in the semiconductor to a conducting region? (J. Appl. Phys.)

9. The dangling bond density in hydrogenated amorphous silicon can be reduced to below $10^{16}\,\text{cm}^{-3}$ but in hydrogenated amorphous SiO_2 it remains above $10^{18}\,\text{cm}^{-3}$. Suggest a reason for this result?

10. How would you increase the electric field breakdown strength of (a) SiO_2 and (b) $BaTiO_3$?

CHAPTER II

Magnetic Properties

Many magnetic properties of thin films are sensitive to structure on levels of scale from the atomic to macrostructure. Advantage has been taken of this dependence on structure to create thin films with properties not observable in bulk solids. However, most magnetic properties do not exhibit the sensitivity to defect structure found for electrical properties and described in Chapter I. Some properties, such as coercivity in soft magnetic materials and the flux line pinning force in high temperature superconductors are affected by the defect structure. Many properties are sensitive to the morphology and dimensions of phases in a multiphase structure.

1. Soft magnetic thin films.

The properties desired from soft magnetic films used as inductive head core materials are large saturation flux density, low coercive field with small hysteresis losses, low magnetostriction (negative λ_s), high initial permeability at high frequencies (i.e. no permeability decrease at high frequencies), small, but non zero, uniaxial anisotropy, low noise and a control over the domain structure in the various portions of the core.

The low coercivity alloy permalloy has a near zero value of the magnetostriction at the composition 81% Ni, 19% Fe and, hence, this composition is that preferred for thin film inductive recording heads, which almost always are under stress in the as-deposited condition. Most of the discussion to follow in this section relates to this material, except as noted. The magnitude of the detected signal increases with the saturation magnetization and permeability. The detection sensitivity of the inductive head increases with decrease in the coercivity and the noise. The magnitude of the detected signal and the head noise are also affected by the magnetic anisotropy and domain structure. Of these parameters, the saturation magnetization alone is insensitive to microstructure in the soft magnetic films.

1.1. Coercivity.

Unfortunately, there is no theoretical basis for predicting the effect of structure upon most of the magnetic properties, although many of them, such as coercivity, are structure sensitive. For example, to be able to predict the effect of

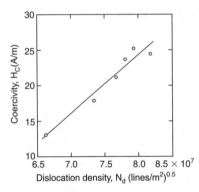

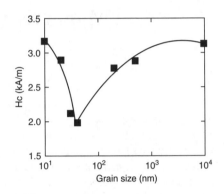

Figure 2.1. (a) An apparent correlation between coercivity and dislocation density deduced from X-ray rms microstrain measurements. (From R.L. Anderson, A. Gangulee and L.T. Romankiw, J. Elect. Mater. _2_, 161(1973) with permission. Copyright 1973 TMS–AIME.)

Figure 2.1. (b) Coercivity of Ni as a function of grain size. (From M.J. Aus, B. Szpunar, U. Erb, G. Palumbo and K.T. Aust, MRS Symp. Proc. <u>318</u>, 39(1994) with permission.)

structure upon coercivity it is necessary to know the mechanism of the domain response to the applied magnetic field. Given this knowledge then an attempt can be made to apply the applicable model for the effect of structure upon coercivity.[1] However, no general rule can be stated that covers all the permalloy thin films in use. Nevertheless, in the regime of nanocrystalline grain sizes it appears that the coercivity decreases with decreasing grain size.[2,3] This correlation is likely to be a manifestation of domain wall boundary motion limitation of the magnetization reversal.[4] Indeed, "Barkhausen noise" manifestations of read instabilities in such materials have been associated with domain wall motion.[5] The correlation found between coercivity and dislocation density shown in Figure 2.1a, which is a manifestation of the barrier to domain wall motion offered by dislocations,[6] is further evidence that domain wall motion governs the coercivity in such nanocrystalline ($Ni_{81}Fe_{19}$) permalloy films. However, in precipitation hardened permalloy containing Nb it has been inferred that magnetization reversal occurs by coherent rotation. If the inference is valid then one would expect that in this material coercivity would vary inversely with grain size.[7] The fact is that this permalloy film has a fine grain, nanocrystalline, grain structure. It is known that the dependence of coercivity on grain size reverses below a certain critical grain size when the domain rotation process becomes thermally activated.[8] Perhaps this situation applies to the latter material. Such a reversal is shown for Ni in Figure 2.1b.

It has been reported that coercivity decreases as the thickness of permalloy films increases.[9] A theory of this effect has been developed by Néel for Bloch wall domain motion[10] which predicts that coercivity should vary as the inverse 4/3 power of the thickness. The experimental value of this exponent is hardly ever $-4/3$. Possible contributions to the deviation of the exponent from the predicted value are the variations of the average grain size and surface roughness with film thickness.

The coercivity of $Ni_{80}Fe_{20}$ seems to be insensitive to texture in that an epitaxial film having (100) texture was found to have the same coercivity as a polycrystalline film of the same thickness having (111) texture.[11]

Two other soft magnetic thin film materials are Sendust (FeAlSi) and amorphous thin films containing either transition metals or magnetic rare earth elements or both as components. Coercivity values in soft magnetic amorphous alloys can be less than 0.01 Oe. The coercivity values in Sendust vary between 0.01 and 0.1 Oe. All soft magnetic thin films must have low values of magnetocrystalline anisotropy energy and magnetostriction. The differences in coercivity between different soft magnetic films may derive from their differences in the resistance to motion of domain boundaries. Amorphous films are likely to have the least resistance to domain wall motion because they are the most homogeneous of the films. Any domain wall pinning centers in the amorphous film are present in a dense population and exert a near net zero pinning effect, i.e. the forces exerted by the pinning centers on opposite sides of the relatively rigid domain walls cancel out. The low coercivity of Sendust alloys is believed to stem from the incompletely long-range-ordered $D0_3$ structure with concomitant weakly pinning grain boundaries, i.e. the disordered regions at the grain boundaries may be present in the same number density as within the grains. It appears that any discontinuity in the structure or sharp gradient in it acts to hinder domain wall motion and thereby increase the coercivity when it is controlled by domain wall motion. An extreme example of the effect of microstructural discontinuity on coercivity in permalloy films is found in films that are line-of-sight deposited in relatively poor vacuum to produce the low density microstructure characteristic of zone 1 deposition (i.e. intercrystalline void network (see I)) in which columnar grains are separated by thin voids. In this case, an in-plane coercivity of 30 Oe was measured.[19,20]

A new soft magnetic alloy under investigation, because it has a higher saturation magnetization than the other soft magnetic materials considered above, is FeMN, where M = Al, Ta.[12–16] However, too few data have been obtained to be able to provide a discussion of the effect of structure on coercivity for this material. To date the investigations on these films have all concentrated on relating properties to deposition parameters.

1.2. Magnetic anisotropy.

The most significant contribution to magnetic anisotropy in $Ni_{80}Fe_{20}$ permalloy films arises from Fe–Fe pair ordering. Such ordering is induced upon

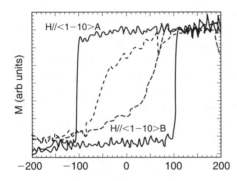

Figure 2.2. Magnetic hysteresis loops for epitaxial $Ni_{80}Fe_{20}$ deposited on (111) Si at 300 K. The field is applied along two different <110> directions in the (111) plane, where A is the axis perpendicular to the in-plane direction of the Ni flux, and B is either one of the two other <110> axes. (From J.R. Childress et al., MRS Symp. Proc. 384, 203(1995) with permission.)

deposition in a magnetic field or upon annealing at a temperature that is high enough to allow the required atomic jumps to take place in bulk. (The latter option is not usually chosen because concomitant grain growth will take place with an increase in coercive force.) Other contributions to magnetic anisotropy may arise from anisotropic stress in the film plane, and from anisotropic distributions of dislocations, stacking faults, twins and grain boundary voids. However, the contribution of internal film stress to magnetic anisotropy in $Ni_{80}Fe_{20}$ permalloy films is likely to be small, because the magnetocrystalline anisotropy constant is controlled mainly by the composition and has a near zero value at the $Ni_{80}Fe_{20}$ composition.

A very interesting effect of a mode of line-of-sight deposition on in-plane magnetic anisotropy has been reported for epitaxial permalloy films deposited under ultra high vacuum conditions.[17] When the substrate is *not* rotated during deposition the cubic unit cell of the permalloy film becomes distorted into a tetragonal one, with compression of the unit cell in the film plane parallel to the projected direction of the incoming flux, for as yet an unknown reason. A consequence of this unit cell distortion is known to be a uniaxial magnetic anisotropy whose amplitude and direction is determined by the magnetoelastic constants of the material with the production of hard and easy axis in-plane magnetic hysteresis loops as illustrated in Figure 2.2. (The magnetocrystalline anisotropy constant of $Ni_{80}Fe_{20}$ is not precisely equal to zero.) The effect just cited is not the same as that observed in polycrystalline permalloy films deposited at pressures between 10^{-7} and 10^{-6} Torr under line-of-sight conditions.[18] Smith, et al.[19] and Cargill et al.[20] have shown that films line-of-sight deposited at pressures higher than 10^{-7} Torr and upon substrates at room temperature (and at less than -60 V bias, in the case of sputtered atoms as the vapor source) produce aligned columnar grains separated by voids having the magnetic easy axis out of the film plane and along the columnar axis.

The existence of some magnetic anisotropy in thin film inductive heads is important because it affects the domain structure and hence the output level of a head, as well as the noise properties. The domain structure in the pole tip region

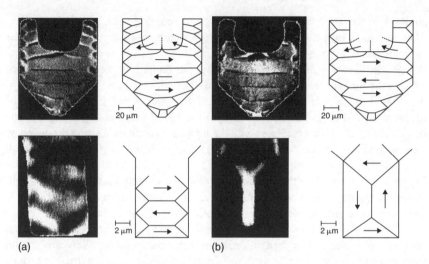

Figure 2.3. Domain images: top–top yoke, bottom–top pole tip. (a) Low noise head with 1.6% read-back amplitude variation, (b) high noise head with 4.6% read-back amplitude variation. (Copyright 1996 International Business Machines Corporation. Reprinted with permission from IBM J. Res. Develop. 40(3).[22])

adjacent to the gap is usually different from that in wide area films of the same thickness. It has been shown that in FeTaN that the domain structure in small pole tip stripes can be changed by varying the saturation magnetostriction constant and the anisotropy field, which is achieved by control of the nitrogen concentration. It is not at this time possible to predict the domain structure in such pole tips. In the case of permalloy heads it has been found that the domain structure in the pole tip is controlled by the relative magnitude of the stresses parallel to the long direction of the pole head and transverse to it and these can be manipulated through control over the local composition and plating conditions.[21] Figure 2.3 illustrates both the desirable and undesirable domains that can appear in permalloy head poles and yokes.

1.3. Noise.

In permalloy inductive heads the origins of write and read noise have been determined. Both write and read instabilities responsible for the noise are associated with small irreversible jumps in domain wall positions which produce abrupt changes in the coil-linked flux.

Such jumps come about as a result of thermal depinning of the domain wall in the case of the write instability. The write operation results in a temperature pulse in the yoke. The most unstable domain walls were found to be spike-like ones near the back gap closure of the yoke in one investigation.[23] Defects likely to pin domain walls are surface scratches, surface roughness, inclusions, and voids. Also, variation in local stresses in films where the magnetostriction constant is non-zero can act to pin domain walls.

In the case of the read instability, the domain configuration that contributes the most to the noise is the 180° wall oriented parallel to the direction of flux flow,[22] in the pole tips and especially in the lower apex region of the top pole tip, as shown in Figure 2.3b, a site where the signal flux density is largest. The magnitude of the noise correlates with the longitudinal magnetization produced by the domain configuration of Figure 2.3b. The larger is the length of the 180° wall in the signal flux flow direction, the higher is the noise amplitude, as shown in Figure 2.4. As noted in the previous section, the domain configuration is controlled by the deposition conditions and the local composition, and experiment is necessary to determine the optimum conditions for any given head design.

Another technique available to reduce noise is to arrange for the head yoke and pole regions to contain only one magnetic domain. There are several techniques proposed to accomplish this objective. One is to make use of a proprietary method,[24] which combines stress with negative magnetostriction. Another involves the phenomenon of exchange pinning in which an antiferromagnetic layer is adjacent to the NiFe ferromagnetic layer, with the spins in the antiferromagnetic layer aligned in one direction. This configuration is used for magnetoresistance based head designs and will be considered in greater detail later in this chapter. Still another involves use of thin ferromagnetic layers interspersed with non-magnetic layers to make up a multilayer head assembly.[25a]

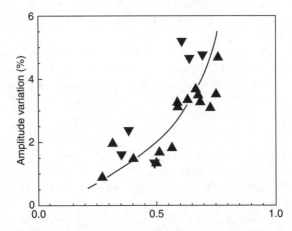

Figure 2.4. Read-back amplitude versus the fraction of longitudinal magnetization. (Copyright 1996 International Business Machines Corporation. Reprinted with permission from IBM J. Res. Devel. 40(3).[22])

1.4. Permeability roll-off.

The decrease in permeability that occurs beyond some frequency of magnetization reversal is a consequence of eddy-current damping. Increasing the resistivity of $Ni_{80}Fe_{20}$ permalloy by decreasing the grain size should improve its resistance to permeability roll-off, but the grain size of permalloy heads is already close to the minimum that can be achieved economically. What is needed is a resistivity that makes the eddy current skin depth larger than the thickness of the pole tips. Thus, other techniques are required. The most effective method of eliminating the decrease in permeability with frequency is to make the magnetic layer thickness less than the eddy current skin depth and to use a multilayer assembly of such magnetic layers separated by insulating non-magnetic ones, with a sufficient number of magnetic layers to obtain the desired flux. Figure 2.5 illustrates the effect of multilayering (curve (b) relative to (a)) as well as of the resistance of the non-magnetic layer (curve (c) relative to (b)) on the frequency roll-off of the initial permeability.

1.5. Soft magnetic spinel type ferrite films.

Recording heads are not the only possible application of soft magnetic thin films. The possible use of integrated circuits in telecommunication devices for operation at extremely high frequencies ($\geqslant 1$ GHz) has inspired investigations of the properties of soft magnetic oxide thin films. The properties required of such thin films are low conductivity (to minimize eddy current losses), a Curie temperature above room temperature, a high saturation magnetization, and minimum magnetostriction. Attention has been focused primarily on spinel type ferrite films. For $NiFe_2O_4$ films[25b] deposited onto (100) $SrTiO_3$ substrates it was found possible to produce single crystals with $(100)_{film}//(100)_{substrate}$ and

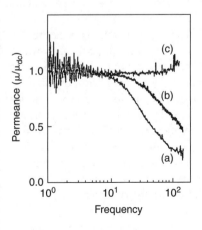

Figure 2.5. Permeance versus frequency. (a) Monolithic $Ni_{80}Fe_{20}$, 1.5 μm thick; (b) laminated film consisting of fifteen 1000 Å thick $Ni_{80}Fe_{20}$ layers separated by fourteen 100 Å thick Zr layers; and (c) laminated film consisting of fifteen 1000 Å thick $Ni_{80}Fe_{20}$ layers separated by fourteen 100 Å thick SiO_2 layers. (From M.A. Russak, C.V. Jahnes, M.E. Re, B.C. Webb and S.M. Mirzamaani, IEEE Trans. Magn. <u>26</u>, 2332(1990) with permission. Copyright 1990 IEEE.)

$[001]_{film}//[001]_{substrate}$. However, the unit cell of the film was tetragonal as compared to the cubic cell of bulk $NiFe_2O_4$. This distortion was due to stress induced in the film either by lattice mismatch or by differential thermal expansion. Despite the monocrystallinity of the film there were inhomogeneities that affected the coercive force and anisotropy field, as indicated by the fact that annealing subsequent to deposition markedly decreased both these parameters. It was speculated that the magnetic inhomogeneities in the film are sub-grain boundaries. However, voids represent an equally likely possible origin of these inhomogeneities.

The saturation magnetization in the $NiFe_2O_4$ thin film has about the same value as in bulk material and is not sensitive to the changes brought about by annealing and which affect the coercivity. Saturation magnetization of $(Mn_{0.46}Zn_{0.54})Fe_2O_4$ films, however, is sensitive to the degree of mismatch of the lattice parameter with the substrate.[25c] This mismatch gave rise in the X-ray rocking curve to FWHM values that differed markedly between a value of 2.5° for a $SrTiO_3$ substrate and a value of 0.9° for a $SrTiO_3$ substrate having a buffer layer of $CoCr_2O_4$ between it and the $(Mn_{0.46}Zn_{0.54})Fe_2O_4$ film. On the latter buffered substrate the saturation magnetization had the bulk value (320 emu/cm³), whereas on the unbuffered substrate the value was no more than 220 emu/cm³. Apparently, identification of the relationships between structure and properties in these soft magnetic ferrite films remains to be accomplished.

2. Hard magnetic thin films.

The properties required of hard magnetic thin films vary according to the application. For in-plane magnetization in magnetic recording type films, the magnetic anisotropy energy, $K_{\perp}$ must be less than the shape anisotropy energy, $2\pi M_s^2$ or the anisotropy field, H_K, must be less than the demagnetizing field, $4\pi M_s$. For magnetization perpendicular to the film plane the reverse relations are required.

Contrary to the case of soft permalloy films, the saturation magnetization, M_s, of a film is sensitive to structure for hard magnetic films. Indeed, as shown by experiments in which ferromagnetic elements are deposited into isolated pores in the shape of columns oriented with the column axis perpendicular to the plane of the film, *the saturation magnetization is proportional to the fraction of pores in the film that are filled with the magnetic material.*[26] This result is significant because for most deposition processes of films at low temperature the film microstructure consists of columnar grains, which are likely to be separated by voids (see I).* The material adjacent to these voids may or may not be oxidized depending upon the vacuum existing during deposition or they may or may not be filled with non-magnetic impurities that are insoluble in the solid state of the deposited material.

---

* "I" refers to Volume I of *Materials Science in Microelectronics*. This book is Volume II in this series.

As an example of the significance of this effect on the saturation magnetization, electrodeposited Co, under conditions which produces a 100% dense, dendritic microstructure without intercrystalline voids, yields a film in which the magnetization vector is in the plane of the film, whereas when the Co is deposited in such a way as to produce a less dense, columnar microstructure, in which the columns are separated by non-magnetic impurities, the magnetization vector is normal to the film plane *even though in both cases the easy magnetic axis [0001] of the Co grains is oriented normal to the film plane.*[27] However, it should be recognized that the transition from longitudinal to perpendicular orientation of the magnetization vector in the cited example is not due solely to the effect of a decrease in the volume fraction of the ferromagnetic constituent. A theoretical treatment of an analogous microstructure by Iwata et al.[28] in 1966 yielded the result that the ratio of $K_\perp/2\pi M_s^2$ increases *strongly* with increase in the ratio of the intercolumnar space thickness to column diameter, i.e. an increase of intercolumnar space relative to that of the columns should increase $K_\perp$. Support for this interpretation is provided by results for Co–M alloys, where M is a non-magnetic element that has limited solid solubility in Co at the usual substrate temperatures. In Figure 2.6, the data indicate that as the saturation magnetization decreases from the value for pure Co (1400 emu/cc) the anisotropy field H_K initially increases. In Figure 2.7, it is shown that an increase in the Cr concentration in Co decreases the saturation magnetization.

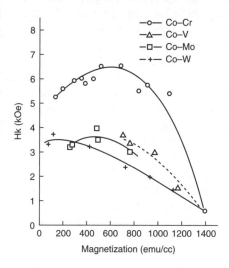

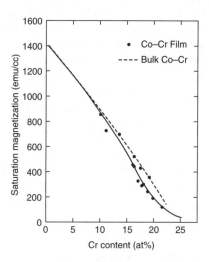

Figure 2.6. Relation between perpendicular anisotropy field and saturation magnetization for various Co–M alloys, where M has limited solid solubility. (From S. Iwasaki, K. Ouchi and N. Honda, IEEE Trans. Magn. MAG-16, 1111 (1980) with permission. Copyright 1980 IEEE.)

Figure 2.7. Effect of Cr concentration in Co base alloy on saturation magnetization. (From S. Iwasaki and K. Ouchi, IEEE Trans. Magn. MAG-14, 849(1978) with permission. Copyright 1978 IEEE.)

(This effect is a consequence of the fact that as the concentration of M increases beyond the solid solubility limit more M appears in the intercolumnar space. Thus, the apparent effect of concentration on both saturation magnetization and anisotropy field is really an effect of microstructure on these parameters. This subject will be discussed in greater detail in Section 2.2.1.)

It would be gratifying to be able to write that this knowledge was used in the development of longitudinal and perpendicular magnetic media. However, as with most developments, understanding followed rather than preceded experiment. Now that we have this knowledge it can be put to use in the further development of these and other magnetic materials. However, not all aspects of microstructure are predictable and in such an event recourse to experiment will be necessary.

2.1. Longitudinal hard magnetic thin films.

The signal output from longitudinal hard magnetic film media is determined[31] by the product, $M_r t$, of the remanent magnetization, M_r, and the film thickness, t. Also, the transition width[32] (directly related to the bit length) scales in some fashion with $M_r t / H_c$, where H_c is the coercivity. Hence, high coercivity and very low film thickness are desired for magnetic media for this application with the remanent magnetization as high as necessary for adequate signal output without sacrificing resolution. Noise is currently the factor limiting the design of these media. Although lower media noise correlates with lower values of coercive squareness,[33] the latter can be changed without any change in media noise,[33] and decreasing the coercive squareness has a deleterious effect on the overwrite performance (the degree to which the residual signal associated with old information is reduced when overwritten by new data). The more significant correlation is between media noise and intergranular exchange coupling – the lower is the latter, the lower is the media noise.[34] Coercivity and media noise are structure sensitive.

Subsequent to the publication of the original edition of this book a review[117] appeared that details the factors involved in the design of hard magnetic materials for longitudinal recording application. Indeed, as described in this review, media noise is the limiting factor in the design of the longitudinal hard magnetic thin film media. As detailed in the Appendix the use of antiferromagnetic layers to enhance thermal stability has succeeded in enabling the production of longitudinal media providing 100 Gbits/in.[2] areal density.

2.1.1. Co alloy thin films.

2.1.1.1. Easy-axis orientation.

Modern longitudinal media consist of Co alloys on Cr substrates.[35] Early in the development of these films it was found that the Cr underlayer assured the in-plane orientation of the easy magnetic axis of the Co alloy film. (In the absence of a Cr layer, or any other underlayer involving epitaxy with the Co alloy overlayer, and in the absence of special deposition techniques, most hcp Co based films tend to deposit with the c-axis normal to the film plane.) It was not until later in the development that the epitaxial effect of the Cr underlayer was understood. Depending upon deposition method and conditions the crystallographic texture of Cr films tends to be either (001), (112) or (110). (see I.) The epitaxies between the hexagonal lattice of the Co based alloys and the bcc Cr underlayer are as follows: for (100)Cr, the (11$\bar{2}$0)Co layer is parallel to Cr(100) with the [0002]Co parallel to [01$\bar{1}$]Cr or [011]Cr;[36] for (211)Cr, the epitaxial relationship is (10$\bar{1}$0)Co parallel to (211)Cr and [0002]Co parallel to [01$\bar{1}$]Cr;[37] for (110)Cr, the easy axis [0002]Co is 28° out of the Cr plane with (10$\bar{1}$1)Co parallel to (110)Cr.[38]

Thus, in-plane magnetization is assured via use of a Cr underlayer. However, since the out-of plane component of the easy magnetic axis orientation varies between these textures the magnetic properties of the Co alloy films will vary also. Further, there is no control over the in-plane azimuthal orientation for any of the Cr underlayer crystallographic textures when the Cr layer is polycrystalline. Also, the thicker the magnetic layer the smaller is the fraction of grains having an epitaxial relationship to the underlayer. Hence, not all Co alloy grains are oriented to provide the optimum signal in present day longitudinal disk media.

The problem of how to process the magnetic and epitaxial underlayers so as to achieve a circumferential in-plane orientation for the magnetic easy axis of all the Co alloy grains is of current concern. One possible method of achieving some control over the in-plane easy-axis orientation is via circumferential mechanical texturing (i.e. the production of controlled surface roughness in the Ni–P/Al–Mg substrate in a circumferential pattern).[39]

2.1.1.2. Coercivity.

The Cr underlayer also serves the function of increasing the in-plane coercivity of the composite film. It is believed that this result stems mainly from the effect of the Cr underlayer on the in-plane orientation of the easy axis in the Co layer. Indeed, increasing loss of the epitaxial relation with increase in the Co layer thickness (i.e. greater percentage out-of-plane orientation of the magnetic easy axis) was found to bring about a loss in the coercivity.[40] (However, it must be mentioned that others believe that the main effect of a Cr underlayer on the coercivity stems from its effect in producing higher energy gradients in the magnetic layer, perhaps corresponding to fluctuations in magnetocrystalline anisotropy, internal stress, and local exchange.[41]) For very thin Cr underlayers, the coercivity increases with thickness.[42] Hence, the effect of Cr layer thickness on coercivity is first to increase it and then to decrease it when the texture degrades beyond some critical thickness.

The coercivity is also a function of the magnetocrystalline anisotropy and the microstructure. For strongly oriented films ((11$\bar{2}$0)Co parallel to Cr(100)) on single crystal underlayers, with a c/2 plane spacing in the Co overlayer initially smaller than the spacing (a$\sqrt{2}$) along the <110> Cr unit cell direction, increase of the a lattice parameter of the underlayer by alloying with Ti resulted in a peaking of the magnetocrystalline anisotropy energy and the coercivity at the composition that yielded equality in these spacings, i.e. minimum epitaxy induced film stress.[43] It has been reported[44] that the magnetocrystalline anisotropy of Co alloys increases with increase in the volume of the unit cell, as, for example by alloying the Co with Pt, Ta and B. However, it is not known whether this effect is a direct one or an indirect one due to the effect on film stress via the mechanism just described. It is apparent that there is still a need for further investigation of the effect of microstructure on coercivity in the Co/Cr longitudinal hard magnetic thin films.

2.1.1.3. Media noise.

Reduction of the exchange coupling between grains significantly enhances the coercivity and reduces the media noise.[34] Such reduction in exchange coupling between grains can be achieved via separation of the grains by empty space or by a paramagnetic layer. The former condition is achieved in deposition at temperatures below T_1, as described in detail in I. The latter condition can be achieved via phase separation or spinodal decomposition, both processes occurring at the film surface during deposition. (Again see I for a more complete description of these processes.) The thickness of the Cr underlayer can affect the grain size of the magnetic layer and the separation between the grains of the magnetic overlayer. In particular, the thinner is the Cr layer the finer is the grain size of the overlayer, the smaller is the separation between the grains of the latter layer and the larger is the intergranular exchange coupling.[45] Thus, a thick Cr layer is desired to reduce the latter parameter. However, there is a limit to the Cr layer thickness above which undesirable effects on the magnetic properties of the magnetic thin films override the reduction of noise that is achieved.[46] Further, from the viewpoint of signal-to-noise ratio, as detailed below, the objective is to reduce grain size rather than increase it via the use of a thicker Cr layer. Hence, techniques of achieving a decrease in the exchange coupling between grains, other than that of increasing the separation between grains, are desired. Among these techniques are: partial oxidation of films deposited in zone 1, which can form oxide layers along the grain boundaries of the magnetic layer[47] thereby providing a paramagnetic layer between the magnetic grains; and the use of insoluble non-magnetic species in the Co alloy, such as Cr,[48] Pt,[49] Ta,[50] and B,[51] which can be induced to concentrate along the grain boundaries and act to separate the grains or to decompose spinodally within the grains or both.

For Co alloys that include Cr as a constituent a behavior diametrically opposite to that detailed in the previous paragraph for the effect of Cr underlayer thickness on the media noise has been reported.[52] In this case care was taken to deposit the thin films under extremely clean conditions, i.e. a sputtering apparatus in which the base pressure is less than $3 \cdot 10^{-9}$ Torr and the impurity level in the sputtering medium (Ar)

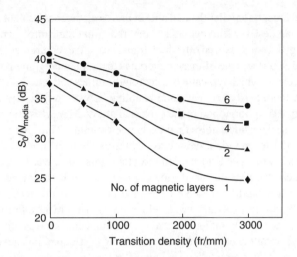

Figure 2.8. Ratio of isolated pulse signal to noise versus transition density for films containing 1, 2, 4, and 6 layers. (From S.E. Lambert et al., IEEE Trans. Magn. <u>26</u>, 2706(1990) with permission. Copyright 1990 IEEE.)

is less than 1 ppb (H_2O) at the point of use. It was found possible to separate the effects of intergranular exchange coupling and grain size on signal-to-noise ratio. The results were that at a given value of the intergranular exchange coupling the signal-to-noise ratio decreased with increasing grain size and that at a given grain size the signal-to-noise ratio decreased with increasing intergranular exchange coupling.

The media noise in these films increases with the recording density.[53] Since a minimum number of grains is required per bit volume to achieve an adequate signal to noise ratio (i.e. at least several hundred grains per bit) then an increase in the recording density requires a decrease in the grain size. As noted above, the media noise is decreased primarily by reducing the ferromagnetic exchange coupling between grains,[52] as by the techniques described in the previous paragraphs, and by decreasing the grain size.[52,42] Also, decrease in the magnetic film thickness can increase the signal to noise ratio.[52,54] Further decrease in the media noise can be achieved by the use of multiple magnetic layers separated by non-magnetic layers,[55] as illustrated in Figure 2.8.

Thus, in summary, there is a marked effect of microstructure on the magnetic properties of Co based longitudinal hard magnetic films, as evidenced by the effects of texture, grain size, the mode of physical separation of the grains, epitaxy induced stress, magnetic film thickness and multilayering of the magnetic medium.

2.1.2. Other hard longitudinal magnetic thin film media.

The need for higher density recording inspires the search for other longitudinal hard magnetic media. For media to have areal densities in excess of 10 Gb/in², the grain size must be smaller than 10 nm. In this case, the requirement for magnetic stability in a single-layer film demands[56] that the ratio of anisotropy energy, $K_u V$ to thermal energy kT must exceed 60, which translates to the requirement that the coercivity exceed 3000 Oe. One of these materials currently undergoing investigation is barium hexaferrite, which typically has coercivity values between 3000 and 4000 Oe and in which there exists magnetostatic coupling between adjacent grains, in place of the exchange coupling found in the Co based alloys. Magnetostatic coupling usually entails lesser transition noise than exchange coupling. Thus, barium hexaferrite has the potential to provide a denser recording medium than do Co based alloys. However, much work needs to be done to attain this objective. Because the same effects of microstructure on the magnetic properties in both magnetic media are expected then it should be possible to use the knowledge already gained with Co based alloy media to expedite the development of barium hexaferrite, and other potential super dense longitudinal hard magnetic thin films such as SmCo and CoPt, i.e. use a polycrystalline underlayer having a strong crystallographic texture to maximize in-plane easy axis orientation via epitaxy, refine grain size, separate grains, etc.

2.2. Perpendicular hard magnetic thin films.

The same magnetic properties required for the longitudinal hard magnetic thin films and which were described in the previous section are required for the perpendicular thin films, with the exception that the easy magnetic axis must lie perpendicular to the film plane rather than lying in-plane.

Achievement of the orientation with the easy magnetic axis perpendicular to the film plane is accomplished either by epitaxy with an underlayer or by appropriate deposition. Co tends to deposit with its [0001] axis normal to the plane of the film. However, to obtain a perpendicular magnetic state it is necessary for the anisotropy field H_K to exceed the demagnetizing field $4\pi M_s$. This situation does not occur for pure, dense Co and it is necessary to alloy the Co or to decrease the volume fraction of Co by physical separation of the Co columns in order to decrease the saturation magnetization, M_s, and increase the anisotropy field. Cr when dissolved in Co acts in such a manner. Thus, Co based alloys containing a concentration of Cr higher than about 18% are the favored magnetic media for perpendicular recording.[30] The underlayer, when used, is hcp Ti, oriented with its c-axis perpendicular to the film plane.

2.2.1. Co–Cr alloy films.

Magnetic isolation of each grain by means of Cr segregation to the grain boundaries is achieved with a suitable substrate temperature that allows the Cr to surface diffuse to the grain boundaries during deposition. The Cr concentration in the Co matrix is increased to above about 26 a% to achieve a sufficient separation of the grains. Segregation is not limited only to the intercolumnar regions, but upon appropriate deposition can be induced to occur within the columns thus leading to the development of Cr poor and Cr rich stripes both perpendicular to the columnar boundaries and containing the columnar axis (and magnetic easy axis direction) and having a separation between neighboring stripes of 3 to 7 nm.[57] As noted, the Cr segregation occurs at the film surface during deposition. Since, the Cr rich area contains about 30 a% Cr it is paramagnetic (26 a% Cr is the limiting composition), while the Cr poor area (about 7 a% Cr) is ferromagnetic.[58] Thus, when these stripes are present the size of the average magnetic unit is smaller than the columnar grain. Stripes are advantageous from the point of view of minimizing media noise because they increase the number of magnetic units in a bit. The magnetic isolation achieved by Cr segregation acts to enhance the coercivity.

It is worth dwelling upon the fact that measurements reveal that the actual Cr content of the ferromagnetic regions is about 7 a%. According to Figure 2.7 and conventional wisdom this amount of Cr in Co would not yield perpendicular magnetization, i.e. 13 a% Cr is the composition below which and above which longitudinal and perpendicular magnetization films are obtained, respectively. Indeed, these observations of ferromagnetic Cr poor regions and non-magnetic Cr rich regions confirm the proposal made in Section 2 that saturation magnetization and anisotropy field are sensitive to microstructure and that the latter affects the orientation of the film's magnetization vector.

These films in which the Cr content is segregated may be expected to contain internal stress with its attendant effect on coercivity. This writer is not aware of any investigation of this effect in perpendicular Co–Cr alloy films. Also, given the epitaxial relationship between the Ti underlayer and the magnetic layer, control of the grain size of the former layer should provide control over the grain size of the latter layer. Finer grain size is preferred to decrease media noise for a given signal level.

The mode of magnetization reversal in fine grained perpendicular type magnetic media containing sufficient Cr to produce the stripes mentioned above is a mixture of rotation and domain nucleation in the soft magnetic initial layer that has a longitudinal orientation for its easy axis. Analysis of hysteresis loops indicates that the average magnetic unit is smaller than one column.[57]

2.2.2. Other perpendicular hard magnetic thin films.

2.2.2.1. Easy-axis and magnetization vector orientation.

The easy-axis perpendicular orientation for the Co–Cr films of the previous section is achieved by epitaxy to an underlayer having a preferred orientation. Perpendicular orientation of the magnetic easy axis can be achieved by a variety of other mechanisms. For example, many hcp Co alloys tend to deposit with the [0001] (the easy magnetic axis) perpendicular to the plane of the film and in this orientation the surface energy is minimized. (See I for discussion of origin of texture in thin films.) This result was found, for examples for a CoNiReP film deposited on a NiMoP underlayer via electroless-plating. The NiMoP underlayer had a random distribution of its grain orientations and, hence, had no effect on the texture of the CoNiReP overlayer which was an outcome of the growth process.[59] There are many examples of textures resulting from such an effect in the literature.

However, to achieve perpendicular orientation of the magnetization vector it is usually necessary to have another contribution to the anisotropy field. As mentioned previously, Co can have a perpendicular orientation of its magnetic easy axis yet yield a longitudinal orientation of the magnetization in the absence of a columnar structure or when the columnar structure is characterized by a column length to diameter ratio about unity. When the column length to diameter ratio greatly exceeds unity and when there is an intercolumnar void network or a non-magnetic material between the columns then there is a shape anisotropy contribution to the anisotropy field that may be sufficient to make the magnetization vector be perpendicular to the field. Another example, where the shape anisotropy contributes to the perpendicular magnetization is illustrated by the case of iron needles which were deposited in pores (tunnels) in an anodized aluminum film, where the pores are perpendicular to the film plane, by electrolysis in an iron sulfide bath. The iron grows as a single crystal with the [110] axis parallel to the needle axis. The [110] direction is not the easy axis direction. The latter lies at 45° to the needle axis. The iron moment lies about 15° from the needle axis and this direction is determined by competition between the demagnetization field of the iron needle, the magnetic anisotropy field of the iron crystal and the dipole interactions with the other needles in the array.[60]

Although there are other contributions to the anisotropy field besides the magnetocrystalline anisotropy and the shape anisotropy, these two parameters, as described above are responsible for the choice between in-plane and perpendicular orientation of the magnetization vector of most crystalline films used in magnetic recording at this writing. The only exception for films consisting of more than a few monolayers may be found when there exists a preferential nearest-neighbor ordering of the magnetic constituent. One such material is Co–Pt. In this system long range ordered phases appear at equilibrium in the phase diagram. Although films are generally deposited in a metastable state it is possible through the use of somewhat elevated substrate temperatures during deposition to produce long-range order (lro) in such films. For example, lro was found in $Co_{50}Pt_{50}$ films deposited at 500°C. When deposited on MgO, epitaxy yielded the equilibrium CoPt fct structure with the c-axis, the easy magnetic axis, normal to the film plane, and a film with perpendicular magnetization.[61]

Very thin films consisting of a few monolayers may involve still another contribution to the anisotropy field that overwhelms the others. In this case, a uniaxial anisotropy resulting from the reduced symmetry at interfaces may overcome the magneto-static energy $2\pi M_s^2$ of the film to yield a perpendicular magnetization.[62] Also, a magnetoelastic contribution to the anisotropy due to epitaxial stress in very thin films may overwhelm the other contributions to determine the nature of the magnetization. These very thin films are of interest in magneto-optical recording and we will return to a consideration of them later in this book.

2.2.2.2. Coercivity.

Coercivity is microstructure sensitive. For example, in the case of Fe needles deposited in pores produced by anodization of Al, as shown in Figure 2.9 the coercivity was found to depend upon the needle diameter for a given porosity. It is apparent that both the perpendicular and parallel coercivity decreases with increase in the needle diameter, except for Ni needles,[26] for which the coercivity first increases then decreases with needle diameter. Coercivity variation with substrate temperature and film thickness in Co–Cr alloys have been correlated to variation in the strength of the [00.2] c-axis texture.[63] However, in the case of the effect of substrate temperature there may also be a contribution due to a decrease in the ferromagnetic column diameter due to increased segregation of the Cr with increasing substrate temperature. Further, the reduction of the ferromagnetic exchange coupling between columns is believed to result in an increase in the coercivity.[57]

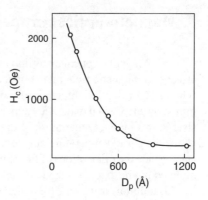

Figure 2.9. Dependence of perpendicular coercive force for iron columns located in pores of an anodized aluminum film on the pore diameter. (From M. Shiraki et al., IEEE Trans. Magn. MAG-21, 465(1985) with permission. Copyright 1985 IEEE.)

Other contributions to the anisotropy energy may also affect the coercivity. For example, it was reported that the larger coercivity of a CoO/Fe_2O_3 multilayer assembly on a NiO substrate as compared to that on a glass substrate was due to the larger anisotropy energy of the former due to a tensile stress in the film plane of the former assembly that was absent in the latter one.[64] Incidentally, these films exhibited perpendicular magnetization with properties suggesting potential use as a high density perpendicular recording medium.

2.2.2.3. Media noise.

The considerations applicable to longitudinal magnetic media with respect to media noise appear to be applicable to perpendicular magnetic media as well, i.e. reduction of exchange coupling between grains by segregating nonmagnetic material in the grain boundaries or within grains by spinodal decomposition, decrease in grain size, etc.

3. Magneto-optical properties.

Magneto-optical recording makes use of the Kerr effect, which is an effect of a magnetic field on the optical polarization of a transmitted beam. In this type of recording the direction of the magnetization vector in a domain is determined by an external magnetic field during the write operation, which usually consists in raising the temperature of the domain by means of a focussed laser pulse above the Curie temperature and cooling in the presence of the external magnetic field. Since the orientations of the remaining non-illuminated domains must not be affected by the external magnetic field it is necessary to have a sufficient coercivity. In addition, to have a high signal to noise ratio the Kerr rotation angle, θ, should be as large as possible. Further, an appropriate temperature dependence of the saturation magnetization and the coercive force, and a low Curie temperature must be achievable. Usually it is also desired that the magneto-optical film have perpendicular magnetization, a square M–H loop and low noise.

Through the use of certain amorphous magnetic media, which are uniform on the scale of 50 Å, noise is at a minimum and not a factor requiring further attention. In such media, the Kerr angle θ is small, $0.3°$, but usable for longer wavelengths than blue.* Such media have as constituents, Fe, Co and R, where R = Tb, Gd or Dy, and are ferrimagnetic. Because of the latter property, the magnetic properties of magnetization, anisotropy, and coercivity can be changed almost continuously by changing R. The only effect of structure in such amorphous films is that of pair ordering, which responds to the magnetic field on cooling below the Curie temperature, and determines the magnetization vector. Indeed, as a result of the work of Harris and associates[65] we may be secure in the knowledge that such pair ordering occurs and affects properties. They found, through the use of polarized synchrotron radiation and EXAFS, that more Fe–Fe in-plane nearest-neighbor pairs existed than out-of-plane ones and more Fe–Tb out-of-plane nearest-neighbor pairs existed than in-plane ones when there was perpendicular magnetization. Also, they found that in the absence of magnetic anisotropy there was a random distribution of these pairs. Compressive stress acts as a driving force to reorient Fe–Fe pairs to be in-plane and Fe–Tb pairs to be out-of-plane because such reorientation leads to a decrease in the stress-free in-plane area and, hence, to a decrease in the compressive stress. Of course, annealing is required to provide the mobility for such pair reorientation.

Structure does affect the properties of potential crystalline replacements for the rare earth-transition metal amorphous alloy magnetic media. Such replacements

---

* However, recent work suggests the possibility that by use of a Ag reflector layer in place of the current Al alloy one the figure of merit at blue wavelength can be improved. Further, it was found necessary to control the smoothness of the interfaces with the TbFeCo layer to obtain maximum signal to noise ratio and elimination of domain boundary pinning.[108]

may be necessary if the Kerr rotation angle for the UV spectrum cannot be improved. It is barely at a usable value now. One candidate system consists of either a multilayer assembly of Co and Pt layers or codeposited films of either CoPt or $CoPt_3$. The improvement in the Kerr signal at blue wavelengths relative to TbFeCo for both these CoPt type films is shown in Figure 2.10.

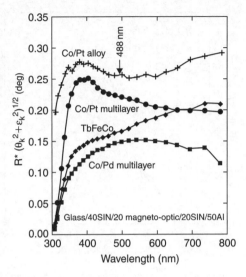

The interface contribution to the perpendicular anisotropy of Co/Pt multilayers may have two origins. One is intrinsic and was first suggested by Neel,[66] while the other depends upon the degree of mixing at an interface and the degree of long range order of any superlattice formed in the mixed region.[67] The intrinsic term is orientation dependent in Co/Pt,[68] with the value

Figure 2.10. Kerr signal as a function of wavelength for various magneto-optic materials. (From T. Suzuki, C.-J. Lin and A.E. Bell, IEEE Trans. Magn. 24, 2452(1989) with permission. Copyright 1989 IEEE.)

for the case where both Co and Pt layers have (111) orientation about twice that for the case where both layers have either (110) or (100) orientations. Indeed, for the same set of multilayer parameters and with the Co layer thickness ≥ 5 Å, use of the (100) orientation yields a longitudinal magnetization, whereas use of the (111) orientation produces a perpendicular magnetization.[69] Understandably, the fraction of alloyed material present increases with increasing substrate temperature, with decreasing bilayer period, and with increasing energy of atoms impinging on the growth surface.[70] Coercivity and effective perpendicular anisotropy appear to increase with increase in the perfection of the (111) texture and smoothness of a SiN substrate. However, on a ZnO substrate the coercivity is independent of the latter parameters and the effective anisotropy constant is nearly twice the value as on a SiN substrate while the half width of the (111) X-ray diffraction peak for the ZnO substrate is about half that for the SiN substrate. Thus, it appears that the effective anisotropy constant increases with the perfection of the (111) texture, while the coercivity is really independent of it and of the smoothness of the substrate, at least in this multilayer assembly consisting of 9 bilayers of 17 Å Pt/5Å Co.[71a] The dependence of anisotropy energy on the perfection of the (111) texture is confirmed by the work of Li and Garcia.[71b]

Reference to Figure 2.10 shows that the alloy films have a still higher Kerr signal than do the Co/Pt multilayer films over the wavelength spectrum investigated

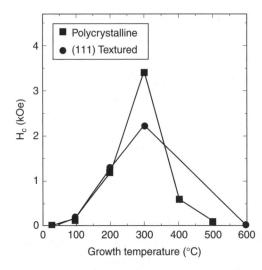

Figure 2.11. Coercivity versus substrate temperature for polycrystalline and epitaxial CoPt₃ films. (From E.E. Marinero et al., MRS Symp. Proc. 313, 677(1993) with permission.)

from blue to near infra-red. This result has been confirmed by other investigators.[72] Thus, the alloy films have been subject to intense investigation.

For co-evaporated films at the CoPt₃ composition grown onto a variety of substrates at a series of increasing substrate temperatures Marinero et al.[73] found that for all films there was a progressive increase in both perpendicular anisotropy and coercivity with substrate temperature up to 300°C above which both properties collapse with increasing temperature and the anisotropy becomes in-plane above 500°C. Included in the substrates was one which yielded a (111) epitaxial film. As shown in Figure 2.11 the polycrystalline films had higher maximum coercivity than the epitaxial film. Further, the coercivity and perpendicular anisotropy varied between the different polycrystalline films. The grain size was the same for all the polycrystalline films and, hence, could not be responsible for the difference in magnetic properties between them. Further, no correlation between magnetic properties and film stress was obtained. The belief was stated that the magnetic properties were dependent upon the parameters that define short-range-order, i.e. the population of Co–Co pairs and their orientation. Magnetization reversal in a (111) CoPt₃ film involves both domain nucleation and domain wall propagation. Athermal, as well as thermal activated magnetization processes contribute to the reversal mechanism with the coercivity predominantly determined by pinning on grain boundaries.

Another ordered CoPt alloy occurs at the 50% composition and has the $L1_0$ structure. This ordered phase only appears upon annealing of the film above about 550°C,[74] with the c-axis oriented perpendicular to the film plane.[75] The coercive force correlates to the fraction of the $L1_0$ phase in the film and the Kerr rotation angle at the 633 nm wavelength is 0.4 for a film annealed at 675°C for 14 hours. The grain size in both types of ordered films is on the order of 5 nm as deposited, and grows to a thickness limited grain size. This result may imply some contribution to media noise for the larger grain size films.

There are other contenders for a magneto-optical recording medium that have better properties at short wavelength than the amorphous TbFeCo alloy.

However, at this writing their likelihood of being accepted commercially is not sufficient to justify a detailed discussion of the effect of structure on their properties in this volume. Suffice it to say that the most obvious effect of structure in the $(BiDy)_3(FeGa)_5O_{12}$ candidate is that of grain size on media noise and in the MnBi candidate of a phase transition and of the grain size, again on media noise.[76]

4. Magnetoresistance.

The magnetoresistance of materials exhibits three ranges of magnitude. In increasing order they are the normal, giant (GMR) and colossal magnetoresistant (CMR) effects. The magnitudes of the change in resistance ($\Delta R/R$) are 2–3% (at 4–5 Oe), 10–20% ($\sim$10 Oe) and >70% (70 kOe), respectively. In addition, $\Delta R/R > 70\%$ has been found in tunnel junctions between ferromagnetic materials, and has been denoted as the tunneling magnetoresistance effect (TMR).

4.1. Normal magnetoresistant films.

Permalloy has been the material used in read heads of disk recording devices. The materials requirements for such read heads are large magnetoresistance, small magnetization, uniaxial anisotropy, small hard axis coercivity, low magnetostriction, low resistivity, wear and corrosion resistance. The sensitivity to the detection of a signal is proportional to $\Delta\rho/\rho$ and inversely proportional to the effective anisotropy field. Uniaxial anisotropy is required to control the domain structure. Small magnetization is required to obtain a low demagnetization field and a large signal sensitivity. A small hard axis coercivity confers a small magnetoresistant hysteresis. Low magnetostriction is to limit stress induced anisotropy to a minimum value. Finally, the lower the resistivity the higher is $\Delta\rho/\rho$. The properties that are most structure sensitive are $\Delta\rho/\rho$ and coercivity. In particular, $\Delta\rho/\rho$ increases and coercivity decreases as the grain size decreases. However, the change in $\Delta\rho/\rho$ is no more than about 1% for a significant change in grain size.

In ferromagnetic materials the resistance can decrease initially with increase in the magnetic field and then will increase. The initial decrease in resistivity is related to the domain structure in the same way that it is related to it for the giant magnetoresistance materials. The origin of this effect will be discussed in the section on the giant magnetoresistance materials. The increase in resistance is caused by a lengthening of the electron path due to the effect of the magnetic field in the direction of the current, which brings about a helical motion in place of the linear motion between collisions. The need for a linear response to the magnetic field, as well as the need to reduce head noise, has led to the requirement that the

MR head material consist of a single domain. Various schemes involving use of adjacent layers have been developed to produce such single domain structures. One such type of adjacent layer makes use of an antiferromagnetic material that is exchange coupled to the ferromagnetic permalloy film.[77] Another makes use of a permanent magnet in the form of a CoPt film.[78]

4.2. Giant magnetoresistance.

There are many different arrangements of magnetic, and non-magnetic, metallically conducting materials that exhibit giant magnetoresistance (GMR). In all these arrangements, the giant magnetoresistance effect stems from the scattering of electrons of one spin state by electrons having another spin state. In contrast to the increase in resistance that comprises the normal magnetoresistance, there is a decrease in resistance with onset of a magnetic field in the GMR effect, as illustrated in Figure 2.12. In this figure the resistance prior to application of the magnetic field was 54 μΩ. It should be noted that application of the magnetic field resulted in a hysteresis, which suggests that there may be a concomitant change in domain structure following magnetic saturation.

It is useful to refer to certain experimental results to understand the manifestations of GMR and its origin. Let us start with Figure 2.13 which shows the GMR as a function of thickness of the non-magnetic layer separating two magnetic layers for current *perpendicular* to the film plane and for a multilayer film of Co/Cu. We note that the GMR oscillates with increase in the thickness of the Cu film. The explanation

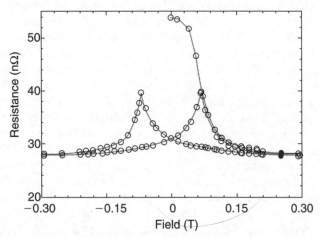

Figure 2.12. Magnetic field dependence of the resistance in a Co/Cu multilayer film with 0.96 nm Cu thickness. (From N.J. List et al., MRS Symp. Proc. 384, 329(1995) with permission.)

for the oscillation in GMR is that an exchange coupling exists between layers such that as the thickness of the non-magnetic layer increases the sense of the coupling oscillates between a ferromagnetic coupling and an anti-ferromagnetic one.

When the Cu layer thickness is smaller than that corresponding to the first peak in Figure 2.13 the spins everywhere are ferromagnetically aligned and if the mean free path of the conduction electrons is larger than the spacing between layers, then electrons sense electrons having only the same spin and the resistance in the absence and presence of a magnetic field larger than that for saturation is unchanged by the magnetic field, i.e. the GMR is zero. When the Cu layer thickness corresponds to the first maximum in GMR, adjacent magnetic layers are anti-ferromagnetically aligned by the coupling field and the resistance in the absence of a magnetic field is larger in this case than in the previous case corresponding to ferromagnetic coupling because the conduction electrons now sense a spin-dependent scattering due to the oppositely aligned spin states in adjacent magnetic layers. In the presence of a magnetic field sufficient to overcome the anti-ferromagnetic coupling, there is no longer a contribution from spin-dependent scattering and the resistance decreases to that characteristic of the ferromagnetic state, the absolute magnitude of the GMR effect, $|\Delta R/R|$, is at a maximum.

The above explanation assumes implicitly that each layer is magnetically homogeneous, but this implicit assumption is not correct. A single layer corresponds to an inhomogeneous magnetic structure. However, for the GMR effect due to a current perpendicular to the film plane the assumption of a homogeneous magnetic structure is nearly correct because the coupling due to one homogeneous magnetic

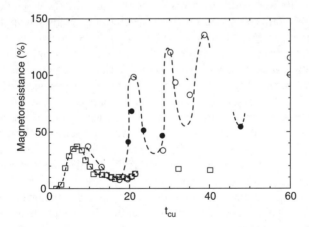

Figure 2.13. Magnetoresistance with current in direction perpendicular to film plane as a function of Cu thickness (in Å). (The squares represent measurements with current in the film plane.) The open circles are for a Co thickness of 2.4 nm and the closed circles for a Co thickness of 1.7 nm. (From N.J. List et al., MRS Symp. Proc. 384, 329(1995) with permission.)

region governs the sense of the spin in the direction normal to the film plane in the adjacent layer. For a current in the film plane the inhomogeneous magnetic structure will negate any GMR effect if the length scale of the magnetic inhomogeneity greatly exceeds the electron mean free path and may eliminate altogether the oscillation in GMR with non-magnetic layer thickness.[79] Thus, we arrive at a possible effect of structure on the GMR, e.g. the effect of the size and presence of magnetic inhomogeneities along in-plane directions in the absence of a magnetic field.

Nothing has been said concerning the identity of the magnetic inhomogeneities present within each layer. In this regard it is worth noting that for many multilayer systems the major source of scattering of spins for in-plane GMR is at the interface between layers.[80] It is believed that the interface regions alternate between regions of intermixing between layers and sharp interface regions involving no mixing.[81] However, to the knowledge of the writer, nothing is known as yet concerning the magnetic structure of the interface regions nor the means of controlling this structure. We do know that when impurities, which exhibit asymmetry with respect to their bulk scattering of spin-up and spin-down electrons, are introduced at Fe/Cr interfaces, then effects on GMR similar to those observed in Fe/Cr multilayers are found if the impurities have the same asymmetry with respect to bulk scattering of spins in Fe as do Cr impurities, or if the asymmetry is inverse to that for Cr in Fe, then the GMR is found to be rapidly degraded.[82] Thus, the spin scattering interface inhomogeneity may well be a region at the interface where atoms of the non-magnetic component are locally dissolved in the magnetic layer.

The above analysis suggests that factors that affect the tendency for solution of two different materials across their common interface should affect the GMR of the multilayer based on these materials. Indeed, systems that exhibit strong interface energy anisotropies with the average bonding between like atoms stronger than between unlike atoms (a positive heat of mixing), exhibit a strong effect of film texture on GMR. For example, such a strong effect is found in the Co/Cu system with (111) texture tending to exhibit much smaller GMR than polycrystalline films.[83] Contrariwise, systems having a negative heat of mixing, such as Fe/Cr, exhibit no effect of different texture on GMR.[84] The degree of mixing at the interface may be dependent upon the relative orientations of the grains adjoining the interface. Hence, the size of the magnetic inhomogeneities may scale with the grain size. Consistent with this analysis is the fact that the in-plane GMR was found to increase in a Co/Cu/Co spin valve as the grain size decreased.[96]

Not only is the spin scattering dependent upon dissolution of non-magnetic impurities into the magnetic layer at the interface, but the phase of the magnetic coupling between magnetic layers is also dependent upon the degree of such interface alloying.[85] This observation may explain why it is found that for a given multilayer structure with non-magnetic layer spacings corresponding to either minima or maxima in the GMR the couplings involve mixed ferromagnetic and antiferromagnetic regions in each layer.[86]

In-plane magnetic inhomogeneity can arise also if the surface roughness is on the order of the layer thickness. Indeed a surface roughness (7 Å rms) has been measured[96] that is on the order of the spacing of Cu corresponding to the first GMR maximum (~ 10 Å) in a Co/Cu/Co spin valve. Consider a sinusoidal roughness wave for such a spin valve, with an amplitude somewhat larger than twice the layer thickness. In this case electrons traveling in the horizontal plane will intersect alternate Co layers each roughness wavelength. Thus, for roughness wavelengths less than the electron mean free path, electrons will be scattered by anti-ferromagnetically aligned spins.

There is a phase transition in Co between the fcc and hcp structures. Co deposits pseudomorphically on Cu in the fcc structure, although the stable low-temperature phase is hcp, and after about 30 Å the remaining Co deposits in the hcp structure. Cu, on the other hand, has the fcc structure in all these multilayer films. Thus, there is the possibility that the first Co layer deposited on a substrate may have the hcp structure, while succeeding layers deposited on Cu have the fcc structure or that Co layers thicker than about 30 Å will consist of both structures. The latter situation has been observed.[87] It has been found in these films that annealing brings about a transition from the hcp to the fcc structure and a reduction in GMR.[87]

For the particular case of a $Ni_{80}Fe_{20}/Cu/Ni_{80}Fe_{20}/Fe_{50}Mn_{50}$ spin valve heterostructure, although a GMR has been measured for both (100) epitaxial and (111) textured films, no GMR was obtained for a polycrystalline film of random orientation.[88] This effect of microstructure on the GMR of a spin valve multilayer has been clarified by Nakatani et al.[89] and is an effect of the crystal structure of the FeMn layer used to pin the magnetization of the adjacent NiFe layer. (The α-Mn (γ-Mn) structure of FeMn is not (is) antiferromagnetic.)

To achieve the low coercivity required for read head GMR devices the solution has been to choose a soft magnetic layer material ($Ni_{80}Fe_{20}$), rather than a hard one (Co). A multilayered NiFe/Cu/NiFe structure is being considered for application in spin valve type read heads, for example. The structure dependence of coercivity in soft magnetic layers has been discussed in Section 1 of this chapter. Spin valves make use of a layer adjacent to one of the magnetic layers to pin the direction of magnetization in the latter, while the magnetization in the other magnetic layer is initially perpendicular to that of the pinned layer, but can change according to the magnetic field it senses. The layer that accomplishes the pinned magnetization is often an antiferromagnet, such as MnFe or an oxide of NiCo, and exchange coupling yields a ferromagnetic alignment in the pinned layer. Several properties of the antiferromagnetic layer, such as the exchange-field strength and the blocking temperature are sensitive to microstructure, and, in particular, to its grain size and crystallographic texture.[89,90]

Giant Magneto Resistance has been observed also in granular films consisting of a mixture of at least two metallic phases, one of which is magnetic and the other not. In this configuration, several magnetic properties including GMR are structure sensitive. In particular, for a mixture consisting of 25% (by volume) NiFe grains and 75% Ag grains, it was found that the GMR first increases then decreases with increase in both the NiFe (and Ag grain sizes), starting from a 23 Å (43 Å) grain size, and with the maximum at a grain size of about 33 Å (100 Å), respectively.[91]

In the case of a film consisting of a mixture of Ag and Co grains containing by atomic fraction 0.33 Co, which was deposited at a sequence of increasing substrate temperatures, it was found that the GMR exhibits a maximum with increasing substrate temperature, the resistivity decreases monotonically with increasing substrate temperature, while the room temperature coercivity increases monotonically with increasing substrate temperature.[92] The latter effect is a consequence of the growth in grain size of the Co particles with increasing substrate temperature, while the decrease in resistivity with increasing substrate temperature is due to the increase in mean free path due to the increase in Ag grain size with increasing substrate temperature. In these mixtures, the ferromagnetic grains are not coupled because their separation is too large. Two opposing tendencies with increasing substrate temperature account for the maximum in GMR. The increase in mean free path tends to increase GMR because an electron samples more Co particles with different spin orientation between scattering events. The increase in Ag and Co grain sizes with increasing substrate temperature, however, causes the electrons to sample fewer Co particles with different spin orientation and, thus, to decrease the GMR.

Another study[93] provided a more concise correlation for the GMR in Co/Ag granular films. In particular, for a given Co grain size the GMR is at a maximum for the smallest Ag spacing between Co particles which provides magnetic isolation for the Co particles and the GMR varies inversely as this spacing. In still another study[94] it was found that, for high concentrations of the ferromagnetic component such that there is ferromagnetic coupling between adjacent magnetic grains, at weak fields there is a positive contribution of the field to the resistance due to anisotropic magnetoresistance. Also, when the components are not completely segregated (metastable solid solutions are still present) there may be a paramagnetic envelope surrounding the ferromagnetic cores as was found for a $(Ni_{81}Fe_{19})_{25}Ag_{75}$ film deposited at room temperature. In this case the magnetoresistance at temperatures above 110°K was found to be dominated by scattering on paramagnetic fluctuations, as in bulk ferromagnets.[95]

Summarizing, in multilayer systems the GMR arises from the presence of spin scattering from regions of oppositely oriented spins, which exist in the absence of a magnetic field, and its absence in the presence of a magnetic field when all the spins have the same orientation. Since electrons sample more oppositely oriented spin regions when the current is perpendicular to the layers (CPP) than when it is in-plane (CIP) the GMR is larger for the CPP measurement. For granular systems, where the magnetic particles are dispersed in a metallic nonmagnetic host and are magnetically uncoupled (randomly oriented spins), and where the dimensions are such that electrons can sample at least several magnetic

particles, the GMR is intermediate to the CPP and CIP of multilayer systems at
their maximum exchange-coupling.

4.3. Colossal magnetoresistance.

Colossal magnetoresistance (CMR) is observed in a variety of systems that
undergo a ferromagnetic to non-ferromagnetic insulator (metal) transition. The
resistivity in these materials is characterized by a peak[97] that occurs at or near the
transition temperature in the absence of a magnetic field, as illustrated in Figure 2.14
for the case of $La_{0.67}Ca_{0.33}MnO_{3-\delta}$. Upon application of a magnetic field in the Tesla
range the resistivity in the ferromagnetic structure is markedly reduced and the

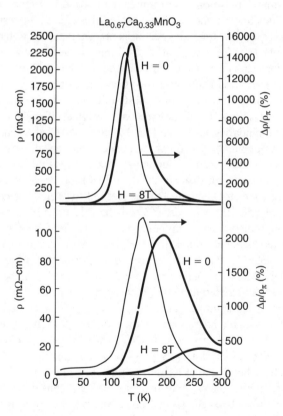

Figure 2.14. The temperature dependence of resistivity for zero magnetic field and an
applied field of 8 T for two $La_{0.67}Ca_{0.33}MnO_{3-d}$ films annealed under different conditions.
(Reprinted with permission from C.L. Canedy et al., J. Appl. Phys. <u>79</u>, 4546(1996).
Copyright 1996 American Institute of Physics.)

temperature of the phase transition is increased. This behavior is characteristic of the systems that exhibit CMR. Thus, in the process of inducing a CMR effect a magnetic field acts to induce the transition from a non-ferromagnetic state to a ferromagnetic state. Hence, it is reasonable to state that the CMR effect is structure dependent.

Not only is the transition temperature sensitive to a magnetic field, but it is also sensitive to applied stress[98] and X-ray exposure[99] in some of the materials that undergo a transition to the ferromagnetic state. These results are consistent with the concept that, in the materials that exhibit CMR, there is a coupling between the spins, the lattice and the charge states.[101] There is evidence also that the material just below the transition temperature is not magnetically homogeneous in zero magnetic field[102a] nor for that matter is it homogeneous just above the transition temperature.[102b] Fluctuations of phase are to be expected in the vicinity of a phase transition temperature. If the observed inhomogeneities are phase fluctuations they should be of interest to investigators interested in this discipline. However, no investigations with this objective appear to have been made for the CMR phase transitions as of this writing. Aside from the implications of these observations with respect to the mechanism(s) of the CMR effect they also suggest additional extrinsic structural effects on CMR. In particular, non-uniform elastic strain,[103] such as that emanating from substrate constraint, or structural imperfections, such as dislocations,[104] may be expected to exert an effect on the CMR.

Grain boundaries have been found to affect the CMR in many manganite systems.[105] Although a CMR is observed in such polycrystalline films, a peak, if present, is much broader than in monocrystalline films of the same composition. Further, and of greater significance, a CMR effect is observed at higher temperatures including room temperature that is induced by a much smaller magnetic field, on the order of 100 Oersteds.[106] The latter effect has been ascribed to spin tunneling through disordered grain boundaries. Indeed, the degree of order at the grain boundaries has been found to correlate to the magnitude of the tunneling magneto resistance (TMR) effect. Coherent grain boundaries were found to yield a much smaller TMR than incoherent grain boundaries.[107] It is believed that the TMR effect occurs in materials in which the spins of carriers are 100% polarized as a consequence of a splitting in the band for the spin carriers with a gap at the Fermi level.[108a] Grain boundaries are not the only barriers being investigated in the TMR effect. Various dielectric layers, such as the intrinsic $(La,Sr)_2O_2$ blocks in layered manganites of the $La_{2-2x}Sr_{1+2x}Mn_2O_7$ compositions,[108b] and extrinsic $SrTiO_3$ layers have been used as barriers in tunneling junctions.[108c]

The final possible effect of structure on CMR considered here is that due to thickness.[109] It has been shown that this effect is not only an indirect effect of strain, but it also is related to an effect on the saturation magnetization. Apparently, the magnetic moment per Mn atom is only 2.5 μ_B in thin films, whereas it can reach 4 μ_B in bulk material. It was suggested that thin films have increasing amounts of antiferromagnetically coupled Mn atoms the thinner the film.

5. Flux line pinning in superconducting thin films.

Defects exert a significant effect on the critical current density in the presence of a magnetic field. They accomplish their effect microscopically by the pinning of the normal cores of superconducting vortices. The defects producing such pinning must be discrete so that there is no interruption in the superconducting current. Thus, defect lines and small defect clusters, but not planar defects except for twin boundaries, can be potentially useful pinning centers in the high T_C superconducting oxides with coherence lengths on the order of 10 Å. The pinning force in the oxide based high temperature superconductors is weak relative to its value in Nb_3Sn. Nevertheless, effort has been expended to discover efficient pinning defects in the high temperature superconductors. One study[110] to determine whether a-axis oriented fine grains in an otherwise c-axis oriented thin film could act as pinning centers was not conclusive. On the other hand, defects produced by high energy particle irradiation, that are localized about the path followed by the energetic particle in the material, were found to increase the pinning sufficiently to allow a hundredfold increase in the critical current density at a magnetic field of 9 kOe.[111] However, the dose of the order of $10^{16}/cm^2$ indicates that the defects are likely to be homogeneously distributed rather than localized about lines. This supposition is supported by transmission electron microscopy of neutron irradiated samples.[112] Dislocations are believed to be flux pinning sites because the core dimension is about the same as the coherence length in the high temperature superconductors. However, the writer is not aware of any experiments that prove that dislocations act as flux pinning sites in high temperature superconductors. Part of the problem in this regard is the production of a controlled array of dislocations lying with their lines parallel to the c-axis. It is possible to obtain dislocations that bound stacking faults that lie in the a–b plane.[113] However, their efficacy in flux pinning is not known and merely suspected. In addition to the line type defects, oxygen vacancies[116] and clusters of point defects are believed to be flux pinning centers also. Similarly, precipitates are believed to act as flux pinning centers.[114] It appears that regions having Ba/Nd < 1.66, of nanometer size, in a single crystal of $Nd(Ba,Nd)_2Cu_3O_{7-\delta}$ having Ba/Nd > 1.66 act as flux pinning centers allowing a J_C of 70,600 A/cm^2 at 77°K and 1 Tesla.[115]

Recapitulation

In soft magnetic thin films coercivity varies with grain size, dislocation density, film thickness and microstructure. For normal permalloy thin films the coercivity depends on the resistance to domain wall motion. Domain walls tend to be pinned by dislocations, grain boundaries, sharp gradients of stress, second phases, voids, surface roughness, and other discontinuities and sharp gradients in the structure. Magnetic anisotropy in soft magnetic thin films depends primarily upon anisotropy in the distribution of the near-neighbor-pair orientation of magnetic atoms and secondarily upon anisotropic distribution of defects, such as dislocations, stacking faults, twins, voids and stress. Noise in soft magnetic thin films is produced by sudden changes in magnetic domain configurations, as occasioned, for example by thermal depinning of domain wall motion. Permeability roll-off is due to eddy current losses which can be reduced by multilayering thinner films.

In hard magnetic thin films saturation magnetization is a function of structure and, in particular, of the intercolumnar space in films deposited in zone 1 or of the unfilled columnar pore density in films produced by deposition into a film with columnar pores. The magnetization vector and the easy magnetic axis are not necessarily colinear. If the microstructure is columnar and the easy axis is out the plane of the film as in pure Co, then the magnetization vector can be changed from in-plane to out-of-plane by increasing the intercolumnar non-magnetic space. This change increases the ratio of anisotropy field to saturation magnetization. The effect of structure on coercivity in longitudinal hard magnetic films is not completely understood although the effect of texture on it is understandable. To optimize signal to noise ratio in longitudinal hard magnetic thin films decrease the grain size and decrease the exchange coupling. The latter can be accomplished by increasing the non-magnetic space between magnetic grains. Also, decrease in the magnetic film thickness and multilayering, alternating magnetic layers with non-magnetic ones, will increase the signal to noise ratio. Decrease of the exchange coupling by increase in the intercolumnar spacing is accomplished by increase in the non-magnetic component of Co base alloys beyond their solubility limit. This procedure and use of underlayer texture to produce an easy axis orientation that is perpendicular to the film plane results in hard magnetic films with a perpendicular magnetization vector.

The population and orientation of pairs of magnetic atoms affects magnetic anisotropy in amorphous rare earth-transition metal alloys and the Kerr rotation angle, coercivity and magnetic anisotropy in long-range ordered alloys of CoPt and $CoPt_3$.

Giant magneto resistance can be induced by alternately multilayering of ferromagnetic and non-magnetic materials with the thickness of the latter such as to induce antiferromagnetic arrangement of spins in alternate ferromagnetic layers by exchange coupling. When the electron mean free path is larger than the layer thickness electrons are spin scattered. The antiferromagnetic arrangement of spins can be changed to the ferromagnetic one by application of a sufficient magnetic field thereby decreasing the resistance. Mixing at the interface between layers can induce magnetic inhomogeneities within a given layer and affect the in-plane GMR if the scale of the inhomogeneity is less than the electron mean free path. GMR can occur also in granular films which contain ferromagnetic grains dispersed in a non-magnetic matrix. The GMR is affected by grain size and volume fraction of the ferromagnetic grains.

Colossa magnetoresistance occurs in manganites that undergo a ferromagnetic to insulator (metal) transition. The resistivity in the absence of a magnetic field peaks near to the transition temperature. The application of a magnetic field decreases the resistivity and shifts the transition temperature to higher temperature. Thus, the CMR effect is intimately associated with a structural transition from a ferromagnetic phase to a non-ferromagnetic one. Grain boundaries introduce another effect known as spin tunneling magnetoresistivity. Other effects of structure on CMR are due to stress and thickness.

Finally, defects act to pin flux lines in superconductors. It is not known how to optimize this flux pinning by defects.

References

1. H. Zijlstra, in Ferromagnetic Materials, Vol. 3, ed. E.P. Wohlfarth, North-Holland, Amsterdam, 1983, p. 37.
2. R.L. Anderson, A. Gangulee and L.T. Romankiw, J. Elect. Mater. 2, 161(1973).
3. T. Jagielinski, MRS Bull. 15, 39(1990).
4. H. Hoffman, IEEE Trans. Magn. Mag-9, 17(1973).
5. A. Chiu, I. Croll, D.E. Heim, R.E. Jones Jr., P. Kasiraj, K.B. Klaassen, C.D. Mee and R.G. Simmons, IBM J. Res. Develop. 40, 283(1996).
6. H. Träuble, in Moderne Probleme der Metallphysik, Vol. 2, ed. A. Seeger, Springer-Verlag, Berlin, 1966, p. 372.
7. E. Adler and H. Pfeiffer, IEEE Trans. Magn. MAG-10, 172(1974).
8. C.P. Bean and J.D. Livingston, J. Appl. Phys. 30, 1208(1959).
9. K.Y. Ahn, J. Appl. Phys. 37, 1481(1966).
10. L. Neel, J. Phys. Radium 17, 250(1956).
11. I. Hashim and H.A. Atwater, MRS Symp. Proc. 313, 363(1993).
12. M. Naoe and S. Nakagawa, J. Appl. Phys. 79, 5015(1996).
13. S. Nakagawa, S. Tanaka, K. Suemitsu and M. Naoe, J. Appl. Phys. 79, 5156(1996).
14. W.C. Chang, D.C. Wu, J.C. lin and C.J. Chen, J. Appl. Phys. 79, 5159(1996).
15. M.H. Kryder, S. Wang and K. Rock, J. Appl. Phys. 73, 6212(1993).
16. Y. Li, B. Cai, X. Zeng and D. Xu, J. Appl. Phys. 79, 4998(1996).
17. J.R. Childress, O. Durand, F. Nguyen van Dau, P.J. Galtier, R. Bisaro and A. Schuhl, MRS Symp. Proc. 384, 203(1995).
18. D.O. Smith, J. Appl. Phys. 30, 264S(1959); T.G. Knorr and R.W. Hoffman, Phys. Rev. 113, 1039(1959).
19. D.O. Smith, M.S. Cohen and G.P. Weiss, J. Appl. Phys. 31, 1755(1960).
20. G.S. Cargill III, S.R. Herd, W.E. Krull and K.Y. Ahn, IEEE Trans. Magn. MAG-15, 1821(1979).
21. P.V. Koeppe, M.E. Re and M.H. Kryder, IEEE Trans. Magn. 28, 71(1992); M. Hanazono, S. Narishige, S. Hara, K. Mitsuoka, K. Kawakami, Y. Sugita, S. Kuwatsuka, T. Kobayashi, M. Ohura and Y. Tsiji, J. Appl. Phys. 61, 4157(1987).

22. A. Chiu, I. Croll, D.E. Heim, R.E. Jones Jr., P. Kasiraj, K.B. Klaassen, C.D. Mee and R.G. Simmons, IBM J. Res. Develop. 40, 283(1996).

23. F.H. Lin, P. Ryan, X. Shi and M.H. Kryder, IEEE Trans. Magn. 28, 2100(1992).

24. R.D. Hempstead and J.B. Money, US Patent 4,242,710 (1980).

25. (a) J.-P. Lazzari and I. Melnick, IEEE Trans. Magn. MAG-7, 146(1971); D. Augier and J.-P. Lazzari, IEEE Trans. Magn. MAG-7, 679(1971); K. Mitsuoka, S. Sudo, M. Sano, K. Nishioka, S. Narishige and Y. Sugita, IEEE Trans. Magn. 24, 2823(1988); J.L. Su, M.-M. Chen, J. Lo and R.E. Lee, J. Appl. Phys. 63, 4020(1988); (b) R.B. van Dover, S. Venzke, E.M. Gyorgy, T. Siegrist, J.M. Phillips, J.H. Marshall and R.J. Felder, MRS Symp. Proc. 341, 41(1994); (c) Y. Suzuki, T.B. van Dover, V. Korenivski, D. Werder, C.H. Chen, R.J. Felder and J.M. Phillips, MRS Symp. Proc. 401, 473(1996).

26. M. Shiraki, Y. Wakui, T. Tokushima and N. Tsuya, IEEE Trans. Magn. MAG-21, 1465(1985).

27. T. Chen and P. Cavallotti, Appl. Phys. Lett. 41, 206(1982).

28. T. Iwata, R.J. Prosen and B.E. Gran, J. Appl. Phys. 37, 1285(1966).

29. S. Iwasaki, K. Ouchi and N. Honda, IEEE Trans. Magn. MAG-16, 1111(1980).

30. S. Iwasaki and K. Ouchi, IEEE Trans. Magn. MAG-14, 849(1978).

31. T. Chen and T. Yamashita, IEEE Trans. Magn. 24, 2700(1988).

32. T. Chen, IEEE Trans. Magn. MAG-17, 1181(1981).

33. I.L. Sanders, J.K. Howard, S.E. Lambert and T. Yogi, J. Appl. Phys. 65, 1234(1989).

34. J. Zhu and H.N. Bertram, J. Appl. Phys. 63, 3248(1988); IEEE Trans. Magn. 24, 2706(1988).

35. J.H. Judy, MRS Bull. XV, 63(1990).

36. J. Daval and D. Randet, IEEE Trans. Magn. 6, 768(1970).

37. T. Hikosaka and R. Nishikawa, IEEE Trans. J. Magn. Jpn. 6, 678(1991).

38. Y. Shen, D.E. Laughlin and D.N. Lambeth, J. Appl. Phys. 76, 8167(1994).

39. M.F. Doerner and R.L. White, MRS Bull. 21, 28(1996).

40. W.T. Maloney, IEEE Trans. Magn. MAG-15, 1135(1979); Y. Hsu, J.M. Sivertsen and J.H. Judy, Proc. PMRC, J. Mag. Soc. Jpn 13 (Suppl. No. S1), 651(1989).

41. T. Yeh, J.M. Sivertsen and J.H. Judy, MRS Symp. Proc. 232, 15(1991).

42. Y. Matsuda, Y. Yahisa, J. Inagaki, E. Fujita, A. Ishikawa and Y. Hosoe, J. Appl. Phys. 79, 5351(1996).

43. N. Inaba, A. Nakamura, T. Yamamoto, Y. Hosoe and M. Futamoto, J. Appl. Phys. 79, 5354(1996).

44. T. Shimatsu, S. Yokota, D.D. Djayaprawira, M. Takahashi and T. Wakiyama, IEEE Trans. J. Magn. Jpn. 9, 34(1994); R.D. Fisher, J.C. Allan and J.L. Pressesky, IEEE Trans. Magn. 22, 352(1986); N. Tani, T. Takahashi, M. Hashimoto, M. Ishikawa, Y. Ota and K. Nakamura, ibid. 27, 4736(1991).

45. T. Yogi, G.L. Gorman, C. Hwang, M.A. Kakalec and S.E. Lambert, IEEE Trans. Magn. 24, 2727(1988).

46. K.E. Johnson, C.M. Mate, J.A. Merz, R.L. White and A.W. Wu, IBM J. Res. Develop. 40, 40(1966).

47. T. Shimuzu, Y. Ikeda and S. Takayama, IEEE Trans Magn. 28, 3102(1992); A. Murayama, M. Miyamura and S. Kondoh, J. Appl. Phys. 76, 5361(1994).

48. J.N. Chapman, I.R. McFadyen and J.P.C. Bernards, J. Magn. Magnet. Mater. 62, 359(1986); D.J. Rogers, J.N. Chapman, J.P.C. Bernards and S.B. Luitjens, IEEE Trans.

Magn. 25, 4180(1989); M.R. Kim, S. Guruswamy and K.E. Johnson, IEEE Trans. Magn. 29, 3673(1993).

49. B.R. Natarajan snf, E.S. Murdock, IEEE Trans. Magn. 24, 2724(1988).

50. D.J. Sellmyer, D. Wang and J.A. Christner, J. Appl. Phys. 67, 4710(1990).

51. C.R. Paik, I. Suzuki, N. Tani, M. Ishidawa, T. Ota and K. Nakamura, IEEE Trans. Magn. 28, 3084(1992).

52. A. Kikuchi, S. Kawakita, J. Nakai, T. Shimatsu and M. Takahashi, J. Appl. Phys. 79, 5339(1996).

53. R.A. Baugh, E.S. Murdoch and B.R. Natarajan, IEEE Trans. Magn. MAG-19, 1722(1983).

54. I.L. Sanders, T. Yogi, J.K. Howard, S.E. Lambert, G.L. Gorman and C. Huang, IEEE Trans. Magn. 25, 3869(1989).

55. S.E. Lambert, J.K. Howard and I.L. Sanders, IEEE Trans. Magn. 26, 2706(1990); E.S. Murdock, V.R. Natarajan and R.G. Walmsley, IEEE Trans. Magn. 26, 2706(1990).

56. P.L. Lu and S.H. Charap, IEEE Trans. Magn. 30, 4230(1994).

57. J.C. Lodder, MRS Bull. XX, 59(1995).

58. K. Yoshida, H. Kakibayashi and H. Yasuoka, MRS Symp. Proc. 232, 47(1991).

59. T. Osaka, T. Homma, K. Noda and H. Asai, MRS Symp. Proc. 232, 65(1991).

60. N. Tsuya, T. Tokushima, M. Shiraki, Y.Wakui, Y. Saito, H. Nakamura, S. Hayano, A. Furugori and M. Tanaka, IEEE Trans. Magn. MAG-22, 1140(1986).

61. G.R. Harp, D. Weller, T.A. Rabedeau, R.F.C. Farrow and R.F. Marks, MRS Symp. Proc. 313, 493(1993).

62. N.C. Koon, B.T. Jonker, F.A. Volkening, J.J. Krebs and G.A. Prinz, Phys. Rev. Lett. 59, 2463(1987); D. Pescia, M. Stampanoni, G.L. Bona, A. Vaterlaus, R.F. Willis and F. Meier, Phys. Rev. Lett. 59, 933(1987); D.P. Pappas, K.P. Kamper, B.P. Miller, H. Hopster, D.E. Fowler, A.C. Luntz, C.R. Brundle and Z.-X. Shen, J. Appl. Phys. 69, 5209(1991); R. Allenspach and A. Bischof, Phys. Rev. Lett. 69, 3385(1992).

63. N. Honda, S. Yanase, K. Ouchi and S. Iwasaki, J. Appl. Phys. 79, 5362(1996); T.M. Coughlin, J.H. Judy and E.R. Wuori, IEEE Trans. Magn. MAG-17, 3169(1981).

64. T. Doi and K. Tamari, J. Appl. Phys. 79, 4887(1996).

65. V.G. Harris, K.D. Aylesworth, B.N. Das, W.T. Elam and N.C. Koon, J. Alloy. Compd. 181, 431(1992); IEEE Trans. Magn. 28, 2958(1992); Phys. Rev. Lett. 69, 1939(1992); V.G. Harris, F. Hellman, W.T. Elam and N.C. Koon, J. Appl. Phys. 73, 5785(1993).

66. L. Néel, J. Phys. Radium 15, 376(1954).

67. M.F. Toney, R.F.C. Farrow, R.F. Marks, G. Harp, T.A. Rabadeau, MRS Symp. Proc. 263, 237(1992).

68. D. Weller, R.F.C. Farrow, R.F. Marks, G.R. Harp, H. Notarys and G. Gorman, MRS Symp. Proc. 313, 791(1993).

69. C.H. Lee, R.F.C. Farrow, C.J. Lin, E.E. Marinero and C.J. Chien, Phys. Rev. B42, 11384(1990).

70. J.A. Bain, B.M. Clemens and S. Brennan, MRS Symp. Proc. 313, 799(1993).

71. (a) S. Sumi, Y. Kusumoto, Y. Teragaki, K. Torazawa, S. Tsunashima and S. Uchiyama, MRS Symp. Proc. 313, 525(1993); (b) Z.G. Li and P.F. Garcia, J. Appl. Phys. 71, 842(1992).

72. D. Weller, H. Brändle, G. Gorman, C.-J. Lin and H. Notarys, Appl. Phys. Lett. 61, 2726(1992).

73. E.E. Marinero, R.F.C. Farrow, G.R. Harp, R.H. Geiss, J.A. Bain and B. Clemens, MRS Symp. Proc. 313, 677(1993).

74. K. Barmak, R.A. Ristau, K.R. Coffey, M.A. Parker and J.K. Howard, J. Appl. Phys. 79, 5330(1996).

75. R. Sinclair, T.P. Nolan, G.A. Bertero and M.R. Visokay, MRS Symp. Proc. 313, 705(1993).

76. T. Suzuki, MRS Bull. 21, 45(1996).

77. D. Markham and F. Jeffers, Proc. Electrochem Soc., (1989).

78. (a) W.C. Cain, D. Markham and M.N. Kryder, IEEE Trans. Magn. MAG-25, 3695(1989); A.P. Malozemoff, Phys. Rev. B35, 3675(1987); (b) C. Brucker and N. Smith Intermag (1990).

79. N.J. List, W.P. Pratt Jr., M.A. Howson, J. Xu, M.J. Walker, B.J. Hickey and D. Greig, MRS Symp. Proc. 384, 329(1995).

80. X.-G. Zhang and W.H. Butler, ibid., p. 323; S.K.J. Lenczowski, M.A.M. Gijs, R.J.M. van de Veerdonk, J.B. Giesbers and W.J.M. de Jonge, ibid., p. 341.

81. C. Meny, J.P. Jay, P. Panissod, P. Humbert, W.S. Speriosu, H. Lefakis, J.P. Nozieres and B.A. Gurney, MRS Symp. Proc. 313, 289(1993).

82. P. Baumgart, B.A. Gurney, D.R. Wilhoit, T. Nguyen, B. Dieny and V.S. Speriosu, J. Appl. Phys. 69, 4792(1991).

83. R.J. Pollard, M.J. Wilson and P.J. Grundy, MRS Symp. Proc. 384, 365(1995).

84. W.-C. Chiang, D.V. Baxter and Y.-T. Cheng, MRS Symp. Proc. 384, 353(1995).

85. B. Heinrich, J.F Cochran, D. Venus, K. Totland, D. Atlan, S. Govorkov and K. Myrtle, J. Appl. Phys. 79, 4518(1996).

86. S. Zhang and P.M. Levy, MRS Symp. Proc. 231, 255(1992); D. Grieg, M.J. Hall, M.A. Howson, B.J. Hickey, M.J. Walker and J. Xu, MRS Symp. Proc. 313, 3(1993); G.R. Harp, S.S.P. Parkin, R.F.C. Farrow, R.F. Marks, M.F. Toney, Q.H. Lam, T.A. Rabedeau, A. Cebollada and R.J. Savoy, MRS Symp. Proc. 313, 41(1993).

87. M.M.H. Willekins, Th.G.S.M. Rijks, H.J.M. Swagten and W.J.M. de Jonge, MRS Symp. Proc. 384, 391(1995).

88. H.S. Joo, I. Hashim and H.A. Atwater, ibid., p. 409.

89. R. Nakatani, K. Hoshino, S. Noguchi and Y. Sugita, Jpn. J. Appl. Phys. 33, 133(1994).

90. T. Tsang and K. Lee, J. Appl. Phys. 53, 2605(1982).

91. M.A. Parker, K.R. Coffey, T.L. Hylton and J.K. Howard, MRS Symp. Proc. 313, 85(1993).

92. W.Y. Lee, V.R. Deline, G. Gorman, A. Kellock, D. Miller, D. Neiman, R. Savoy, J. Vazquez and R. Beyers, ibid., p. 79.

93. M.B. Stearns and Y. Cheng, ibid., p. 393.

94. B. Dieny, S.R. Teixeira, B. Rodmacq, A. Chamberod, J.B. Genim, S. Aufret, P. Gerard, O. Redon, J. Pierre, R. Ferrer and B. Barbara, ibid., p. 399.

95. P.G. De Gennes and J. Friedel, J. Phys. Chem. Sol. 4, 71(1958).

96. R.D.K. Misra, T. Ha, Y. Kadmon, C.J. Powell, M.D. Stiles, R.D. McMichael and W.F. Egelhoff Jr., MRS Symp. Proc. 384, 373(1995).

97. M.F. Hundley, J.J. Neumeier, R.H. Heffner, Q.X. Jia, X.D. Wu and J.D. Thompson, J. Appl. Phys. 79, 4535(1996).

98. A.J. Millis, T. Darling and A. Migliori, MRS Symp. Proc. 494 (1998).

99. V. Kiryukhin, D. Casa, J.P. Hill, B. Keimer, A. Vigliante, Y. Tomioka and Y. Tokura, Nature 386, 813(1997).

100. C.L. Canedy, K.B. Ibsen, G. Xiao, J.Z. Sun, A. Gupta and W.J. Gallagher, J. Appl. Phys. 79, 4546(1996).

101. R.H. Heffner, MRS Symp. Proc. 494(1998).

102. (a) S.J.L. Billinge et al., Phys. Rev. Lett. 77, 715(1996); J.W. Lynn et al., Bull. Am. Phys. Soc. 41, 529(1996); D. Louca and T. Egami, Bull. Am. Phys. Soc. 41, 636(1996); (b) J.Z. Sun, L. Krusin-Elbaum, A. Gupta, G. Xiao and S.S.P. Parkin, Appl. Phys. Lett. 69, 1002(1996).

103. K.A. Thomas, P.S.I.P.N. de Silva, L.F. Cohen, M. Rajeswari, A. Goyal, T. Venkatesan, N.D. Mathur, M.G. Mathur, J.E. Evetts, R. Hiskes and J.L. MacManus-Driscoll, MRS Symp. Proc. 494 (1998).

104. L. Ryen, E. Olsson, C. Kwon and R. Ramesh, ibid.

105. K.M. Krishnan, A.R. Modak, C.A. Lucas, R. Michel and H.B. Cherry, J. Appl. Phys. 79, 5169(1996); S. Jin, T.H. Tiefel, M. McCormack, R.A. Fastnacht, R. Ramesh and L.H. Chen, Science 264, 1021(1994); M. McCormack, S. Jin, T.H. Tiefel, R.M. Fleming, J.M. Phillips and R. Ramesh, Appl. Phys. Lett. 64, 3045(1994); A. Jin, M. McCormack, T.H. Tiefel and R. Ramesh, J. Appl. Phys. 76, 6929(1994); E.S. Gillman, K.H. Dahmen, S. Watts, X. Yu, S. Wirth, and J.J. Geremans, Pros MRS Symp. Proc. 494(1998).

106. J.Y. Gu, R. Shreekala, M. Rajeswari, T. Venkatesan, R. Ramesh and C. Kwon, MRS Symp. Proc. 494 (1998).

107. T. Venkatesen, MRS Symp. Proc. 494 (1998).

108. (a) S.-W. Cheong and H.Y. Hwang, Proc. MRS Symp. Proc. 494, (1998); (b) T. Kimura, Y. Tomioka, T. Okuda, H. Kuwahara,A. Asamitsu and Y. Tokura, ibid.; (c) C. Kwon, Q.X. Jia, Y. Fan, M.F. Hundley, D.W. Reagor and D.E. Peterson, ibid.; G.Q. Gong, A. Gupta, G. Xiao, P. Lecoeur and T.R. McGuire, Phys. Rev. B54, R3742 (1996).

109. S. Jin et al., Appl. Phys. Lett. 67, 557(1995), S. Freisem, J. Aarts, R. Hendrikx and H.W. Zandbergen, MRS Symp. Proc. 494 (1998).

110. Y. Higashida, Y. Sugawara, K. Michishita, Y. Ikuhara, Y. Kubo, K. sasaki, H. Saka and N. Uno, J. Mater. Res. 12, 455(1997).

111. R.P. van Dover, E.M. Gyorgy, L.F. Schneemeyer, J.W. Mitchell, K.V. Rao, R. Puzniak and J.V. Waszczak, Nature 342, 55(1989).

112. M.A. Kirk, MRS Symp. Proc. 209, 743(1991).

113. R. Ramesh, D.M. Hwang, P. England, T.S. Ravi, C.Y. Chen, A. Inam, B. Dutta, L. Nazar and T. Venkatesan, MRS Symp. Proc. 169, 809(1990).

114. D. Shi, J.G. Chen, U. Welp, M.S. Boley and A. Zangvil, Appl. Phys. Lett. 55, 1354(1989).

115. T. Egi, J.G. Wen, W. Ting, T. Higuchi, S.I. Yoo, K. Kuroda, H. Unoki, M. Murakami and N. Koshizuka, in **Advances in Superconductivity** VIII, Vol. 1, ed(s). H. Hayakawa and Y.Enomoto, Springer-Verlag, Tokyo, 1996, p. 465.

116. E.M. Chudnovsky, Phys. Rev. Lett. 65, 3060(1990).

117. D. Weller and M.F. Doerner, Ann. Rev. Mater. Sci. 30, 611(2000).

Appendix

Developments since 1998 in magnetic thin films.

A1. Soft magnetic thin films.

A1.1. Coercivity.

Permalloy as the basis of low coercivity films for thin film inductive write head application is still attractive. Investigation of the effect of strain produced in a pseudomorphic film of permalloy on coercivity[A1] revealed that coercivity decreases with film thickness in the range of thickness below the critical thickness corresponding to the loss of coherency, i.e. generation of misfit dislocations. This effect of thickness on coercivity is revealed in Figure A2.1. The out-of-plane elastic strain in the film increases with decrease in the thickness. The substrate was MgO(001). The in-plane magnetization hysteresis loops exhibit a fourfold <100> cubic anisotropy.

There are a variety of materials that have low coercivity and other properties making them candidates for inductive write head application. A list of electroplateable materials belonging to this group and their properties is given in Table A1. Heretofore, the head material was chosen for two functions: read and write. However, the read function is now performed by MR heads. The write function now requires a higher saturation

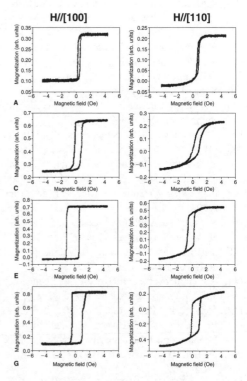

Figure A2.1. Magnetic hysteresis loops (in-plane MOKE) for epi permalloy films with thicknesses (A) 3.5 nm, (C) 10 nm, (E) 35 nm, (G) 100 nm. Critical thickness is about 35 nm. (Reprinted with permission from F. Michelini et al., J. Appl. Phys. 92, 7337(2002). Copyright 2002 American Institute of Physics.)

Table A1.1. Plated materials with high saturation magnetization.

Material	Coercivity H_c (Oe)	Anisotropy field H_K (Oe)	Magnetostriction coefficient λ	Saturation magnetization relative to $Ni_{80}Fe_{20}$
$Ni_{80}Fe_{20}$	0.3	2.5	$-0+$*	1
$Ni_{45}Fe_{55}$	0.4	9.5	$+$	1.6
NiFeCo	1	6–10	$-0+$	0.8–2.4
NiFeCoB	0.6	14		1.5
CoNiFeS	1	20	$+$	1.7
CoFe	3	7–14	$-0+$	1.9
CoFeB	1	7–14	$-0+$	1.9
CoFeCr	0.3	20	$-$	1.7
CoFeNiCr	0.5	$-$	$+$	1.7
CoFeP	1	15	$-0+$	1–1.5
CoFeSnP				
CoFeCu	1	13–18	$-0+$	1.7–2.2
CoB	1	40		1.2
CoFeB (*e*-less)	1	15		1.2

* λ can be negative, zero, or positive depending on alloy composition; specifically for NiFe, λ is negative below about 19 wt.% Fe and positive above about 19 wt.% Fe.

magnetization than that available in the $Ni_{80}Fe_{20}$ composition. Reference to the data in Table A1.1 shows that there are a number of possible candidates to replace the 80/20 permalloy for the write head application. A discussion of them is given in Ref. [A2].

The above are not the only write head materials. Among the other candidates are FeCoZrO,[A3] Co_xC_{1-x},[A4] FeCoN,[A5] FeCoZrBCu,[A6] multilayers,[A7,A8] and others. The problem that needs to be solved is how to obtain the needed higher-saturation magnetization without changing another property, such as magnetostriction coefficient, outside acceptable limits. In these materials the origin of the increase in saturation magnetization is the Fe and Co. The structural factors affecting coercivity are: dispersed particles to minimize grain growth and hence grain size in the FeCoZrO system[A3] and use of soft magnetic layers in a multilayer arrangement with the candidate high B_s layer.[A7–A10] It is believed that texture and rotational symmetry of the texture axis plays a role in the canceling of magnetocrystalline anisotropy energy to a low value in the case where a bcc magnetic-phase grows on a fcc phase with the substrate plane (111). This system belongs to the Kurdjumow-Sachs epilayer group $(110)_{bcc}//(111)_{fcc}$ with $[1,-1,0]//[0,1,-1]$ or one of the other two <110> directions in the (111) plane. For the cancellation of the anisotropy and magnetoelastic energies to take place more than one bcc crystal has to nucleate on one fcc substrate crystal surface.

Another method of achieving a cancellation of the effect of large magnetocrystalline anisotropy and thereby a soft magnetic film is to decrease the dimension of the crystalline magnetic unit to below a critical size and have an amorphous magnetic layer between crystalline units thin enough to transmit the exchange coupling. One example of the application of this principle involves distributions of crystalline particles with high anisotropy in an amorphous matrix.[A6] Another example involved the reduction in grain size of FeCo grains by the multilayering process (reducing the thickness of the FeCo layer) with amorphous Co–M layers between the FeCo layers.[A8] The amorphous Co–M layers act as grain refiners but also as soft magnetic grain boundary material producing exchange coupling between the crystalline layers and thereby enabling the averaging out of the magnetocrystalline anisotropy coefficient K_1 at FeCo layer thicknesses less than about 15 nm. (M = Zr provided the best results.)

It has been shown that the film stress due to a high magnetostriction coefficient also can affect the soft magnetic properties. Only when the film stress was in a limited range of tensile stress was it possible to attain soft magnetic properties in an FeCoB crystalline film.[A11]

A1.2. Permeability roll-off.

Write heads are scheduled to perform at frequencies approaching 1 GHz soon (2005). This level of performance requires that the ferromagnetic resonance frequency (FMR) of the write head film exceed 2 Gb/s. An effort is underway to determine the structural elements that affect the FMR of the new write head materials. The origins of ultra-high frequency losses in permeability are not known as well as for low frequency. However, in Figure A2.2 it is apparent that the roll-off

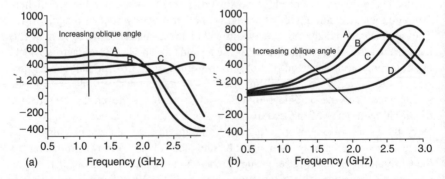

Figure A2.2. Real μ' and imaginary μ'' permeability versus frequency for various angles of line-of-sight deposition. (Reproduced with permission from T.J. Klemmer et al., J. Appl. Phys. **87**, 830(2000). Copyright 2000 American Institute of Physics.)

frequency for an Fe–Co–B amorphous film increases with increasing obliqueness of the line-of-sight deposition angle which implies that the roll-off frequency increases with increasing anisotropy in the amorphous film.[A12] Such anisotropy may have several origins: anisotropic induced stress, anisotropic short range order of the magnetic atoms, and, most likely, columnar growth with intercolumnar voids, i.e. shape anisotropy of the magnetic regions. The FMR for this system is above 2 GHz, as shown in the figure. Eddy-current losses induce damping which act to decrease the FMR and may have other deleterious effects. In FeCoZrBCu multilayer films the damping constant, α, is independent of layer thickness and number of layers.[A6] There is some ambiguity about the desirability of this damping. Some researchers[A5,A6,A12] believe it is desirable to reduce it. Others[A13] believe it may be desirable to increase it.

A2. Hard magnetic thin films.

A2.1. Longitudinal hard magnetic thin films.

At this writing there appears to be a transition taking place to perpendicular recording media. However, because of the capital investment in longitudinal media facilities progress has been made to identify and use new schemes of extending the use of longitudinal media to higher areal bit densities. Among these schemes are the use of antiferromagnetic layers to enhance thermal stability, patterned bits, higher moment write–head coupled with increased media coercivity, single domain particles and magneto-optical techniques.

The structural aspects of the antiferromagnetic concept are: first use of an intermediate non-magnetic layer (Ru) between stabilizing magnetic layer(s) and a main magnetic layer to achieve the antiferromagnetic arrangement as illustrated in Figure A2.3; the use of seed layers to achieve the in-plane orientation of the magnetic layers c-axis (the Ru for the intermediate non-magnetic layer is conditioned by the need for coherency between layers to maintain the needed orientations and the propagation of the grain size through the layers via columnar growth);[A14]

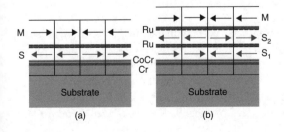

Figure A2.3. Schematic illustration of arrangement of layers in AFC media. (Reproduced from Fujitsu Sci. Tech. J. <u>37</u>, 145(2001) with permission.)

a (11.0) texture for the main magnetic layer produces superior results;[A15] a small grain size for the underlayer enhances the improvement of the antiferromagnetic arrangement.[A15]

The physics underlying the use of antiferromagnetic magnetic coupling scheme is based on two relations. One is that the effective magnetic thickness,

$$(M_R\delta)_{eff} = (M_R\delta')_{mag} - (M_R\delta'')_U$$

is given by the difference between that for the top magnetic layer and the underlying one. Thus, this parameter can be maintained constant at its optimum value while increasing the thickness, δ', of the top magnetic layer. The latter action allows the activation energy of the barrier to thermal instability to be maintained above the minimal limit of about 60. This activation energy is given by the product of the anisotropy constant K and the volume of the magnetic bit. Since increase in the areal density involves a decrease in the diameter of the magnetic grains and the area of the bit, an increase in thickness of the grain allows the volume V to be maintained constant while the areal density is increased. This is achieved with a constant effective magnetic thickness.

Of course, the actual physics is a bit more complicated than indicated in the last paragraph. Nevertheless, the functioning of the AFC to allow an increase of the areal density in longitudinal media to above 100 Gbits/in.^{2} is essentially due to the considerations of the last paragraph. This mode of longitudinal recording is commercial at this writing. In effect, it has stemmed the transition to perpendicular recording media for a while.

None of the other options for extending the areal density of longitudinal recording media have reached the commercial stage at this writing. There are many problems with the single domain particle media concept one of which is achieving alignment of the particles to allow the needed rapid access to each particle. Probably the most advanced of these other options is that of patterning the substrate onto which the single magnetic domains are deposited. At least the patterning problem seems solved.[A16] The AFC has advanced so rapidly because it involves the least change to the mode of manufacturing the disk of all the options.

A2.2. Perpendicular hard magnetic media.

At this writing one commercial perpendicular media recording system has been announced which will start delivery of systems 4/2005 and involves a system with areal density of 133 Gbits/in.^{2} No details were issued concerning this system. However, earlier publications[A17] from the group responsible for this development indicate the possibility that the perpendicular magnetic layer is a CoPtCrO composition coupled with either a Ru underlayer or an FeAlSi underlayer.[A18] Figure A2.4 shows a TEM image of the CoPtCrO layer. Although a scale is not provided with the image it was stated that the grain size was as small as 3–10 nm in diameter. The white area between the dark spots comprise the intergranular space, which is believed to have an amorphous structure. The composition of the intergranular space contains more Cr and O than the grain interiors and also some Co. The dimension of a square bit at 133 Gbits/in^2 is about 70×70 nm^2. Thus, the bit contains many grains, each of which is believed to contain a single domain. It is interesting to compare Figure A2.4 with Figure A2.5. The latter shows a distribution of FePt particles produced by precipitation from solution with a detergent volume concentration of 0.4%.[A19] The size dispersity of the latter is much smaller than for the former with the average size about the same. In this system although not stated in the study[A17] the CoPt grains probably have the $L1_0$ structure with a high magnetic anisotropy.

The $L1_0$-based media have been intensely studied[A20] in view of their applicability both to perpendicular recording as well as to heat assisted magnetic recording. Both continuous and particulate based media have been investigated.

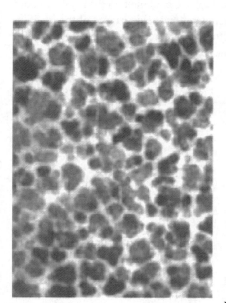

Figure A2.4. TEM bright field image of CoPtCrO magnetic layer. Particle diameter ranges between 3 and 10 nm. (Reproduced from IEEE Trans. Magn. <u>36</u>, 2393(2000). Copyright 2000 IEEE.)

Figure A2.5. SEM of FePt nanoparticles deposited on Si/SiO from a solution containing 10^{-4} parts by volume excess surfactant. (Reproduced from JMMM <u>272–276</u>, Shukla et al., e1349. Copyright 2004 Elsevier.)

The FePt system allows about a 150°C lower transition temperature from the room temperature deposited fcc structure to the $L1_0$ structure than is achievable with the CoPt system. These systems have the advantage of easily forming appropriately oriented particles of very small grain size, <10 nm, which have made them preferable media for these applications relative to barium hexaferrite which at this date (2004) have not yielded grain sizes smaller than about 20 nm.[A21]

A2.3. Magnetic tunneling junction.

The main structural problem with the magnetic tunneling junctions (MTJs) that use amorphous insulator layers is that of providing the thinnest possible pinhole or defect-free insulator layer. This problem need not be confronted in view of the success of the crystalline insulator layer, MgO, in providing much higher tunneling magnetoresistance than heretofore possible.[A22] It seems likely that the MgO-based MTJs with the proper oxidation and annealing processes will bring about a revolution in RAM and other technologies.[A23] It is too early at this writing to provide an overview of the relation between structure and properties for this latest development in spin engineering. However, one comment on the results of both studies in Ref. [A22] is that the highest temperature the interfaces of the junction have experienced prior to test is a controlling parameter on the tunneling magnetoresistance (TMR) produced by these junctions. We show in Figure A2.6 the TMR

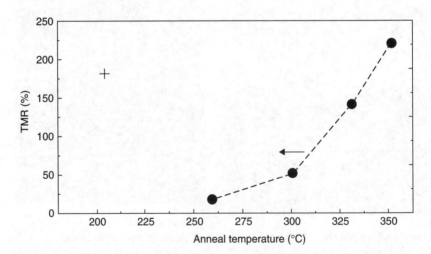

Figure A2.6. TMR as a function of anneal temperature (filled circles) or substrate temperature (+). (Data taken with permission from S.S.P. Parkin et al., Nat Mater <u>3</u>, 862 (2004).)

values achieved by the IBM group upon annealing and also the value obtained by the AIST group on deposition at the substrate temperature of 200°C. The TMR achieved by the latter group on room temperature deposition is 88%. These results taken together suggest to me that a rearrangement takes place at the surface upon deposition at 200°C that does not take place on room temperature deposition. This rearrangement might be the ordering of oxygen vacancies. Thus, the MgO deposited at 200°C has this rearrangement (ordering?) throughout its thickness. However, the MgO deposited at room temperature requires annealing to a higher temperature for this rearrangement to occur because it involves atom jumps in the bulk rather than at the surface. Alternate possibilities are that the elevated temperature is necessary for stress relief or to reduce the dislocation density at the interfaces. Thus, either the structure of the interfaces or the MgO insulator layer affects the TMR. It was suggested in Ref. [A22b] that oxygen vacancies in the MgO affect the TMR. Also, the fact that the tunneling barrier is much smaller than that expected in the absence of defects in MgO suggests that defect control over the tunneling is involved in the mechanism responsible for the TMR observed. It seems likely to me that even higher values of TMR will be found in the future as our understanding of the phenomenon improves. From the observations of spin polarization in these studies one may deduce that spin polarized interactions are responsible for the TMR, but which interactions are really unknown for the studies referenced. The fact that the spin polarization of the electrons transmitted through the MgO layer increased with annealing suggests an effect of structure (interface or defect) on the spin polarization. This is a fascinating development with extraordinary potential.

A2.4. Colossal magnetoresistance.

Many mechanisms for Colossal magnetoresistance CMR have been discovered since publication of I. Here we are concerned with the effect of structure on properties. Also, we will constrain our interest to those materials and their combinations that exhibit a magnetoresistance effect at room temperature, which, after all is the temperature at which most devices using this effect will function. The first is a situation that brings to mind the MTJ device. However, it is accomplished in one material: ferromagnetic $La_{0.67}Sr_{0.33}MnO_3$. The tunneling barrier in this case is a crack (empty space) in the ferromagnetic film.[A24] Actually, many such parallel cracks are produced in the film by having the induced tensile stress exceed the tensile strength. This stress is induced by coherency between the film and substrate. Figure A2.7 reveals an AFM image of the surface showing the cracks. By varying the thickness of the film it is possible to vary the linear density of cracks. Figure A2.8 shows the resistance and relative magnetoresistivity (MR) ratio as a function of temperature for two linear densities of cracks and in the absence of cracks. It is apparent that even at room temperature an MR value of -40% was

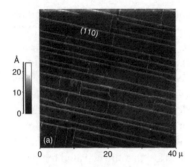

(a)

Figure A2.7. AFM image of surface of $La_{0.67}Sr_{0.33}MnO_3$. (Reproduced with permission from F.C. Zhang et al., Solid State Commun. <u>131</u>, 271. Copyright 2004 Elsevier.)

achieved at a field of only 500 Oe in the film having the highest crack density ($10^4 cm^{-1}$), that the MR value achieves a larger absolute value the higher is the crack density and that in the absence of cracks the MR ratio was close to zero at a field of 500 Oe. According to the analysis in the paper the spin polarization at room temperature was about 0.4. Hence, this system is not as efficient as the MTJ system discussed in the previous subsection. Nevertheless, it suggests that empty space as a tunneling barrier may be a subject that will be investigated in a number of configurations involving the tunneling of electrons having polarized spins. It also implies that empty space is a more efficient barrier than grain boundaries in $La_{0.67}Sr_{0.33}MnO_3$ and perhaps in other materials as well. This development will undoubtedly inspire many studies.

There have been other reports of room temperature CMR, but they are not as large as that just described. In one[A25] the CMR effect arises from grain boundary scattering of the polarized spins in Sr_2FeMoO_6. However, this was observed in a bulk sample with no orientation of the grain boundaries. Perhaps, in a columnar grained thin film the effect may turn out to be larger or smaller depending upon the effect of stress. In the others structure was not investigated except in the macroscopic sense of the device structure, (i.e. p–n junction).[A26]

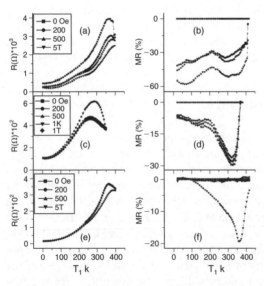

Figure A2.8. (a) and (b) For films having $10^4 cm^{-1}$ crack density, (c) and (d) for films having $2 \cdot 10^3 cm^{-1}$ crack density and (e) and (f) for films having no cracks. (Reproduced with permission from F.C. Zhang et al., Solid State Commun. <u>131</u>, 271. Copyright 2004 Elsevier.)

References to Appendix

A1. F. Michelini et al., J. Appl. Phys. 92, 7337(2002).
A2. P.C. Andricacos and N. Robertson, IBM J. Res. Dev. 42, 671(1998).
A3. S. Ohnuma et al., Appl. Phys. Lett. 82, 946(2003).
A4. H. Wang et al., MRS Symp. Proc. 721, E6.4(2002).
A5. N.X. Sun, A.M. Crawford and S.X. Wang, 721, E6.3.1(2002).
A6. H. Okumura et al., J. Appl. Phys. 93, 6528(2003).
A7. H. Jiang et al., J. Appl. Phys. 91, 6821(2002).
A8. G. Pan and H. Du, J. Appl. Phys. 93, 5498(2003).
A9. N.X. Sun and S.X. Wang, IEEE Trans. Magn. 36, 1506(2000).
A10. H. Katada et al., J. Magn. Soc. Japan 26, 505(2002).
A11. M. K. Minor et al., J. Appl. Phys. 91, 8453(2002).
A12. T.J. Klemmer et al., J. Appl. Phys. 87, 830(2000).
A13. S.E. Russek et al., J. Appl. Phys. 91, 8659(2002).
A14. E.N. Abarra et al., Fujitsu Sci. Tech. J. 37, 145(2001).
A15. K. Tang et al., J. Appl. Phys. 93, 7402(2003).
A16. M. Albrecht et al., IEEE Trans. on Magn. 39, 2323(2003).
A17. S. Oikawa et al., IEEE Trans. on Magn. 36, 2393(2000).
A18. T. Hikosaka et al., IEEE Trans. on Magn. 37, 1586(2001).
A19. N. Shukla, J. Ahner and D. Weller, JMMM 272–276, e11349(2004).
A20. J.-U. Thiele et al., J. Appl. Phys. 91, 6595(2002); Y.K. Takahashi, M. Ohnuma and K. Hono, J. Appl. Phys. 93, 7580(2003); A.C. Sun et al., J. Appl. Phys. 95, 7264(2004); F.E. Spada et al., J. Appl. Phys 94, 5123(2003); and references therein and many others.
A21. A. Morisako et al., JMMM 272–276, 2191(2004).
A22. S.S.P. Parkin et al., Nat Mater 3, 862(2004); S. Yuasa et al., ibid., p. 868.
A23. W.H. Butler and A. Gupta, Nat. Mater. 3, 845(2004).
A24. F.C. Zhang et al., Solid State Commun. 131, 271(2004).
A25. K.-I. Kobayashi et al., Nature 395, 677(1998).
A26. H. Tanaka, J. Zhang and T. Kawai, Phys. Rev. Lett. 88, 027204(2002).

Problems

1. Assume that the free energy of the system decreases when domain walls and grain boundaries are coplanar, although it increases with increasing area of domain walls. Further, the coercivity increases with increasing resistance to domain wall motion. Given this knowledge state show graphically how you would expect the coercivity to depend upon grain size and your reasons for the grain size dependence of coercivity that the curve reveals.

2. Non-magnetic particles that intersect domain walls decrease the domain wall area and, hence, the total free energy. Yet for a given small particle size the coercivity first increases then decreases with increasing number density of particles. Provide an explanation for this effect.

3. What contribution to the magnetic anisotropy described in the text would you suggest is responsible for the magnetic anisotropy in amorphous alloys due to near-neighbor pairs of magnetic atoms? Justify your answer.

4. Estimate the energy of activation for thermal depinning of domain walls that are pinned by: (a) a hexagonal array of grain boundaries; (b) a uniform distribution of spherical particles.

5. In what way does the ratio of anisotropy field to saturation magnetization affect direction of the magnetization vector?

6. Why should texture in hard magnetic thin films affect coercivity?

7. What is the effect of increasing the concentration of non-magnetic elements in a Cr base solid solution beyond the solubility limit on the microstructure?

8. In ideally long range ordered $CoPt_3$, which has cubic symmetry, there are no nearest-neighbor pairs of Co atoms. How do you then explain magnetic anisotropy in this composition material?

9. How do you account for the fact that the magnetoresistance of Co–Cu multi-layers oscillates with increase in thickness of the Cu layers?

10. Define the characteristics of a material which might exhibit a Colossal magnetoresistance effect.

11. In a granular thin film consisting of a mixture of grains of a ferromagnetic material and grains of a non-magnetic material how would you manipulate the microstructure to optimize the giant magnetoresistance effect for that film?

12. What variation of structure affects the Colossal magnetoresistance effect in $La_{0.67}Ca_{0.33}MnO_3$?

13. If you had control over the morphology and distribution of defects in $YBa_2Cu_3O_{7-\delta}$ what defects and how many of them would you use and how would you arrange them to maximize the critical current density at a field of 1 Tesla? What defect would you remove from the thin film to accomplish this objective?

CHAPTER III

Optical Properties

In this chapter we are concerned with the effect of structure on the optical properties of thin films. In the past the optical properties of thin films that were important consisted of the solar and thermal reflectances and transmittances inasmuch as these films were used as passive coatings of lenses, windows, etc. With the advent of the possibility of integration of optical and electrical devices in telecommunication and computing another group of optical properties became significant for thin films, such as luminescence, wave guide transmission, and various non-linear optical properties. Unfortunately, there are few, if any, quantitative theoretical relationships between structure and optical properties. Thus, the discussion to follow is based on qualitative concepts and phenomenology. Nevertheless, the influence of structure on optical properties will be made apparent.

1. Luminescence.

1.1. Effect of lattice parameter on band gap energy.

Structure affects both the wavelength and intensity of the light emitted in the luminescence displayed by various materials. The wavelength of light emitted in devices made from III–V semiconductors corresponds closely to the band gap energy, which, in turn, is dependent on the lattice parameter of the common sphalerite crystal structure, as illustrated in Figure 3.1. As shown, the band-gap energy decreases (i.e. the wavelength of the emitted light increases) as the lattice parameter increases. However, even at a constant lattice parameter there exists a range of band gap energies. Thus, other parameters besides the lattice parameter must affect the band-gap energy.

It has been found that the type of superlattice formed on deposition of thin films affects the band gap energy, without significant change in the lattice parameter. A case in point is the lowering of the band-gap energy (between 50 and 100 meV) upon the onset of growth-induced ordering in GaInP.[1] Another parameter affecting the band-gap energy is strain induced by epitaxy. Thus, multilayers of strained-layer superlattices, with different repeat wavelengths and equal thickness layer components (i.e. different associated strains) have different band-gap energies. These effects of strain on the band-gap energy are used in the band-gap engineering of various devices.

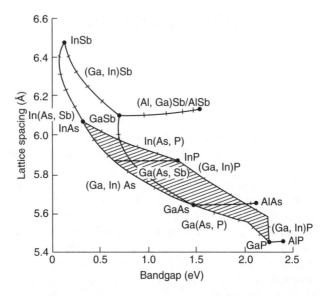

Figure 3.1. Effect of lattice parameter on band gap of III–V semiconductors. (Reprinted from **Concise Encyclopedia of Semiconducting Materials & Related Technologies**, eds. S. Mahajan and L.C. Kimerling, Pergamon Press, Oxford, 1992, p. 441, Copyright 1996, with permission from Elsevier.)

1.2. Effect of defect structure on luminescence degradation.

Other than excited transitions between states of impurity atoms, the mechanism of light emission in a semiconductor involves either radiative recombination of free holes and electrons that are injected into a p–n junction or population inversion in a region near a p–n junction where light is emitted when excited electrons relax back to the ground state. In either case, defects in the structure corresponding to deep-level states can lead to degradation of the optical properties by providing non-radiative centers for hole-electron recombination.

What are the defects that degrade the radiative recombination properties of semiconductors? Obviously, any non-radiative recombination center present lowers, by competition, the intensity of the emitted light. Among such non-radiative recombination centers are regions along extended defects (dislocations, stacking faults, etc.) that act to trap charge, impurity atoms with corresponding deep electron levels, and various other point defects having deep levels. We will not discuss the effects of impurity atoms in detail although they comprise the majority of the deep-level centers provoking non-radiative recombination. However, there are compilations of the energy levels of these defects.[71] In most cases, metal impurities are sources of deep-level non-radiative recombination and usually can be prevented from contaminating thin-film devices by control over deposition and processing. Sometimes non-metallic impurities can act as deep-level non-radiative recombination centers depending upon the host and the impurity. Most of the time these impurities affect the production of these recombination centers indirectly through their effect on the generation of other defects, such as dislocations, or as segregated species at dislocations and interfaces.

Dislocations need not be in the active region at the start of operation of a device, but can climb or glide into these regions from nearby regions during operation and lead to device failure. This type of failure tends to occur in GaAs- and GaP-related materials and II–VI blue-green laser structures. Condensation of excess self-interstitials leads to the formation and growth of 1/2 <100> extrinsic sessile dislocation loops. When the excess interstitials are exhausted the associated degradation will saturate. Heteroepitaxial structures are prone to degradation due to their associated strain which, when the thin film has a thickness close to or above the critical thickness, acts to generate misfit dislocations in or near the interface. Dislocations that move under the influence of stress into the active region of devices originate at such misfit dislocations,[2] at threading dislocations,[3] stacking faults near substrate–epilayer interface,[4] microtwins,[5] vacancy and interstitial loops,[6a] etc. Chu and Nakahara[6b] have provided a defect map for device failure modes in laser diodes. Certain atomic species act to remove the traps at dislocations and grain boundaries that serve as non-radiative recombination centers. For example, unintentional passivation of the non-radiative recombination centers along dislocations by hydrogen during MOCVD of InP has been observed.[54] No adequate explanation at this writing has been given for the observation[7] that dislocations at a density of $10^{10}\,cm^{-2}$ do not degrade luminescence in GaN.*

It is believed that point defects that are generated by non-radiative recombination at some deep-level point defect, can themselves act as deep-level non-radiative recombination centers that generate additional point defects and thusly lead to gradual degradation of the intensity of emitted light in GaAlAs LEDs.[8] The identity of these point defects is not known.[55] The presence of microloops consisting of interstitial-type Frank loops and point-like defects have been observed via TEM after operation, but not before operation of such LEDs. The presence of stress in the LEDs acts to enhance the rates of the reactions associated with the deep-level point defects bringing about more rapid failure of the devices. They occur readily in GaAlAs/GaAs devices, but not in InGaAsP/InP devices. Little is also known as to their control, except the suggestions that strict stoichiometry and careful lattice matching in the heterostructure might ameliorate this

---

* Dislocations may still be sites of non-radiative recombination centers. However, if GaN contains a sufficiently high concentration of radiative isoelectronic complexes that bind excitons then charge carriers may not reach the dislocations, i.e. they are trapped at the radiative complexes. See Chapter VII.

problem. (A possible mechanism for the cloning action of these defects is described in Chapter VII.) Point defects (vacancies?) introduced during low temperature MBE growth, presumably due to low surface adatom mobility, have been shown to be deep-level non-radiative recombination centers in Si and SiGe/Si heterostructures.[9] Positive ion bombardment during such growth enhances the concentration of such defects. Hydrogenation passivates some of these defects, but not all of them.

Interfaces are often the host for deep level defects, such as dangling bonds. These defects act as non-radiative recombination centers and can reduce the intensity of the generated light in semiconductor lasers, which have interfaces adjoining the regions where the radiative recombination takes place. Hydrogen passivation of these defects is a strategy for recovering the radiative recombination properties of these devices.[10] One of the types of interfaces that can be the source of non-radiative recombination centers is that between the ohmic electrodes and the p- and n-layers of LED devices.[2] It is believed that these centers are at dislocations and it is observed that dark spot defects representing concentrations of non-radiative recombination centers grow in size and intensity in the region of high current density suggesting the occurrence of dislocation multiplication and motion during operation of the device.

The defining of point defects in semiconductors is an ongoing objective of research in materials science. Although many defects have been characterized by their energy levels in the band gap, little progress has been made in providing an atomic description of the point defects involved in non-radiative recombination. See Chapter VII for a brief description of what is known microscopically about some of the defects.

1.3. Effect of defects in enhancing luminescence.

Solute atoms need not solely bring about degradation of properties. The luminescence due to donor–acceptor pairs is well known.[56] The interest in Er as a solute in Si* is that it acts to produce Er^{3+} luminescent centers emitting photons at 1.54 μm as a result of an internal transition between an excited state ($^4I_{13/2}$) and the ground state ($^4I_{15/2}$), and thus initiates the hope that a light-emitting device can be developed in Si that can eventually be used in integrated optic applications. By itself, the light intensity emitted by Er at its solubility limit ($10^{18}/cm^3$) is too low to be useful. Thus, use has been made of ion implantation to increase the concentration of Er in the Si beyond the solubility limit and of annealing, without the formation of any erbium silicides, to activate the Er^{3+} ions. Ion implantation involves the likely

---

* The other obvious interest stems from the fact that Si has an indirect band gap and, consequently, the intrinsic recombination efficiency in the absence of impurities is very low.

production of defects in the Si host that can act as non-radiative recombination centers and compete with the Er^{3+} ions for the charge carriers that excite the luminescence. Thus, deposition at a low temperature that hinders diffusion has been used as another strategy to incorporate sufficient Er atoms as solutes.

The problem with Er in Si is that temperature acts to quench the optical emission making the use of this system at room temperature difficult. Since Er is luminescent at room temperature in SiO_2, the effect of Er–O complex formation in Si on the luminescence has been investigated. It has been found that co-doping with oxygen or fluorine, to form electronegative ligands with the Er^{3+} ions, strongly enhances the luminescence.[11] The function of these additional dopants is not really known. It has been suggested that F acts to increase the proportion of the Er present in the optically active Er^{3+} state,[12] that it acts to increase the emission cross-section when adjacent to an Er^{3+} ion,[13] that it acts to passivate point defect non-radiative recombination centers introduced by the ion implantation,[14] or that it helps trap a majority carrier until the arrival of a minority carrier at the Er^{3+} ion since their recombination is necessary to excite the latter to emit light. Obviously, further studies are necessary to clarify this matter and they will undoubtedly be carried out. An analysis of the effect of heat treatment on the light output of a region in silicon containing F and Er atoms suggests that the F atoms are complexed with the Er atoms and that the maximum light output corresponds to an ErF_3 complex.[15a] Recently, luminescence at room temperature has been reported,[15b] but the quantum efficiency obtained was very low. In a film consisting of Er nanoparticles embedded in a poly-Si thin film host, which contains a high concentration of oxygen (~40%), room temperature photoluminescence at 1.54 mm was obtained after a 500 °C anneal.[51] Here, it is also hypothesized that the luminescence is localized at Er–oxygen complexes.

Langer et al.[60] have shown theoretically that Auger de-excitation with energy transfer to free-charge carriers should be one of the main non-radiative quenching processes of the Er^{3+} luminescence. This raises the question as to whether it is possible to excite the luminescence in the relative absence of free carriers. This was first demonstrated[61] for the Er located in the depletion region of a diode using hot electrons in the reverse bias mode with an increase in the room temperature luminescence efficiency. A further increase in this efficiency is potentially achievable through the use of an avalanche diode to provide hot electrons to a greater number of Er atoms located in the carrier free insulating region of the device.[57] The quenching that limits the luminescence due to Er in Si is generally absent in dielectric hosts, another carrier free host material. Thus, Er^{3+} has also been used as a dopant in dielectric wave guides and other dielectric optical media to provide optical gain where this property is useful. For example, a net gain of 4 dB/cm was achieved in an Er-doped soda-lime silicate glass wave guide on Si film.[15c] Quenching of the luminescence due to Er also appears to be absent in porous Si that is subsequently oxidized, in that the quantum efficiency in photoluminescence only decreases by a factor of less than two between 12 K and 300 K.[62]

1.4. Effect of surface area/volume on luminescence in silicon.

Perhaps the most striking effect of structure on luminescence is that associated with porous silicon. The modern interest in porous silicon was nucleated by the work of Canham[16a] and Lehmann and Gösele[16b] who found strong visible room temperature photoluminescence and an increase in the absorption edge by about 0.5 eV, respectively, in porous silicon. SEM and TEM studies show that porous silicon is a coral-like structure, which consists of a continuous hierarchy of columns and pores.[16c] The columns are typically a few tens of nanometers long and a few nanometers in diameter and may undulate in diameter along the length. However, the morphology of the porous silicon structure does not appear to be significant with respect to the existence of the luminescent properties, although changes in morphology can affect details concerning luminescent line width, peak energy, etc.

The origin of the luminescence exhibited by porous silicon is still uncertain. To some investigators the visible luminescence originates from the band-gap enlargement due to quantum confinement in these columns.[17a] To others it involves six member Si rings that are bonded to other structures by oxygen atoms as in siloxene.[17b] To others it is some combination of quantum confinement and surface structures.[17c] The intensity of the photoluminescence scales with the surface area of the porous silicon that is exposed to the environment.[18a] This fact implies that some structure at the surface is involved in the luminescence. However, the defects at the surface that can quench the radiative recombination there must be passivated for this luminescence to be detected. Thus, it has been found that desorption of the hydrogen can remove the luminescence, while rehydrogenation or oxidation can restore the luminescence.[18b] To further substantiate the significance of the surface to the luminescence of porous silicon is the fact that the wavelength of the emitted light is a monotonic function of the ratio of the number of Si–O bonds to Si–H bonds along the porous surface, as revealed by FTIR measurements.[18c] However, it has also been shown that for a given number of particles the photoluminescence peak shifts to shorter wavelength as the particle size decreases in agreement with the expectation from the particle confinement model.[52] Despite this evidence, somewhat more quantitative evidence exists[53] that does not yield any change in the photoluminescence peak wavelength with particle size.

An existence requirement for photoluminescence from porous silicon is the presence of small crystalline silicon units adjacent to anodically etched surface. The charge carriers are generated in the crystalline silicon by light and migrate to the surface where they recombine to produce the luminescence. This fact has been demonstrated by the production of photoluminescence from Si particles that are isolated within an insulating medium.[18d] To develop efficient electroluminescence it appears to be necessary to have the electrode adjacent to the surfaces acting as hosts of the associated radiative recombination centers. If solid electrodes are deposited by line of sight processes it turns out that the resulting luminescence is very inefficient.

If liquid electrolyte is used as an electrode then the luminescence is extremely efficient. One possible reason for this state of affairs is simply that the charge carriers injected by the solid electrode are either trapped or recombined at non-radiative centers before they can reach the radiative centers along the major fraction of the surface removed from the electrode. What appears to be required is a method of depositing a transparent solid electrode, which mimics the electronic behavior of the liquid electrolyte, over all the surface of the porous silicon.

Summarizing, the single structural aspect of porous silicon that relates to the intensity of its luminescent properties is the associated immense surface area per unit volume. The luminescence itself appears to be a property of the crystalline silicon surface having hydrogen and oxygen bonded to the surface.

1.5. Luminescence induced in quantum wells.

One way of increasing the probability that radiative recombination of holes and electrons can occur before non-radiative recombination with defects is to arrange for the distance between holes and electrons to be much smaller than the distance between these charged entities and defects. The confinement of both electrons and holes in a single quantum well and the minimization of the concentration of non-radiative recombination centers in this well is a strategy for achieving this goal.* Electroluminescence in silicon based materials is also a highly desired goal. Both these objectives have been achieved in strained SiGe/Si quantum wells.[50] However, to satisfy the requirements just outlined it is necessary to produce a SiGe/Si structure such that the Si layer, the quantum well layer, does not contain any Ge atoms. Another condition to be satisfied for the achievement of efficient electroluminescence is that the concentration of non-radiative recombination centers be nil. High growth temperature in molecular beam deposition will help attain the latter condition by the elimination of lattice imperfections such as missing atoms and disorder due to atoms displaced from lattice sites. However, high growth temperature also leads to the smearing of the SiGe/Si heterointerfaces. A solution to this problem was found via the use of surfactant layers to prevent the interchange of Ge with Si atoms at the surface during deposition. In the absence of the surfactant, Ge atoms that are just sub-surface interchange with surface Si atoms because in so doing they lower the surface energy. The presence of a surfactant layer removes this driving force for Ge atoms to exchange with Si atoms in the next layer since the latter layer is subsurface beneath the surfactant surface layer and the layer containing the Ge atoms is one layer below the subsurface layer. Using this technique of surfactant assisted growth

---

* This strategy implicity assumes that the non-radiative recombination centers usually present on semiconductor surfaces are absent or passivated. Otherwise, the high surface/volume ratio in these particles would prohibit their use.

SiGe/Si based diodes were made that were electroluminescent at room tempera-ture.[50] Many other types of quantum well structures have been produced that are based on Si including Si nanocrystals imbedded in SiO_2,[63] 4 nm thick polycrystalline Si layers formed by crystallization of a-Si:H layers sandwiched between a-SiN:H lay-ers,[64] and Si nanocrystals embedded in polymeric hosts.[65] Undoubtedly, many more schemes of using quantum well structures to develop electroluminescent devices compatible with silicon technology will be developed subsequent to this book's pub-lication date. As mentioned above the ability to minimize non-radiative recombina-tion in nanocrystalline size units and the property of quantum confinement, or large surface area per unit volume or both, which are also associated with small nanocrys-talline size units stem from the structure-related parameters of size and morphology.

1.6. Luminescence in organic thin films.

Electroluminescence also exists in organic thin films. Two aspects of struc-ture appear to be significant in these materials. One is geometrical in nature in that the use of polymer hole- and electron-injection layers between the electrodes and the luminescent layer endows enhanced properties, in part, because all of the injected charges contribute to radiative recombination without arriving at the electrodes, and there is no possibility of energy transfer from excitons to a metal electrode,[19a] or of non-radiative recombination at defects along this interface, and, in part, because oxy-gen release from an indium–tin oxide electrode is prevented from contributing oxy-gen to the luminescent layer. Dielectric layers, as, for example, provided by 10 Å thick LiF layers between electrodes and active layers, also enhance properties by lowering the onset voltage for luminescence and increasing the current density.[66] The other aspect of structure is associated with light-emitting electrochemical cells denoted as LECs. This aspect is the microstructure of the two phase mixture of luminescent polymer and solid electrolyte. When the morphology of this mixture is that of an interpenetrating two phase network, the solid electrolyte, such as $PEO(LiCF_3SO_3)$, conducts the positive ions and the luminescent polymer, such as PPV, provides the pathway for electron transport.[19b] A p–n junction is formed where the charge carri-ers can recombine radiatively to emit light. Bipolar high-boiling-point surfactants[19c] improve the luminescent properties by reducing the distance between the two phases of the network.

2. Optical transmission in wave guides.

There are several origins of optical losses in wave guide propagation. Among these are absorption, leakage, and scattering. Contributing to absorption losses are defects, both in the grains and at grain boundaries. Impurities and oxygen vacancies in potassium niobate films and strontium barium niobate are among such defects.[20a,b] Leakage is a function of the substrate and cladding layer. For example, for a GaAs substrate, a $LiNbO_3$ film 400 nm thick, at a wavelength of 1000 nm, an MgO cladding layer having a smooth surface would have to be at least one micron-thick in order to limit the loss to 1 dB/cm.[21] There are several sources of scattering: internal, surface, and interface. Internal scattering results from material inhomogeneities. Among the latter are second phases, mixed a-axis/c-axis texture, twinning, grain and domain boundaries. Roughness at surfaces and interfaces contributes to scattering losses. The latter losses peak at some intermediate thickness of the wave guide layer. Surface roughness of the wave guide film depends upon the grain morphology and grain size. It is least for monocrystalline films, highest for films having a columnar grain morphology and intermediate for a fine-grained polygonal morphology. An estimate of the maximum surface roughness to maintain the optical loss less than 1 dB/cm is on the order of 1 nm (RMS). Figure 3.2 shows a plot of RMS roughness of various ferroelectric oxide thin films used for wave guides as a function of film thickness. This result suggests that attention must be paid to film growth conditions to maintain the roughness at

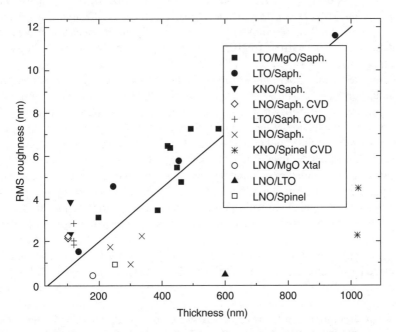

Figure 3.2. Dependence of RMS surface roughness on thickness of polycrystalline ferroelectric oxide thin films. (From Fork et al., MRS Bull. 21, 57(1996) with permission.)

acceptable levels for the film thickness required. There is some hope that such conditions can be met.[22] One possible solution to the surface roughness problem is planarization of the wave guide surface subsequent to growth.

From the above discussion it may be concluded that monocrystalline films are among the most likely options to satisfy the requirements for low optical loss wave guides that can be integrated with Si-based devices. In fact, this conclusion is validated by the results of experiments. For example, Schwyn-Thöny[23] has succeeded in growing (001) oriented monocrystalline films of $KNbO_3$ on oriented spinel and MgO substrates by planar rf-sputter deposition which have optical losses of 1.1 dB/cm at a wavelength of 632.8 nm and an RMS surface roughness of 5 nm. Wessells et al.[24] have also succeeded in growing (001) oriented $KNbO_3$ monocrystalline films on perovskite ($SrTiO_3$) and spinel ($MgAl_2O_4$) substrates having RMS surface roughnesses of 2.7 and 1.2 nm, respectively, which were used to prepare wave guides in which second harmonic generation was observed. For both investigations the monocrystals were multi-domained. Wessels et al., by poling subsequent to film deposition, were able to align these domains and achieve higher non-linear optical properties. Also, although single crystal films of $PbTiO_3$ can be produced on a variety of substrates, the domain structure and the refractive indices resulting depend upon the substrate.[27] McKee et al.[28] found it necessary to deposit the TiO_2 layer of the perovskite structure first during MBE of $BaTiO_3$ and $SrTiO_3$ films on (001) MgO in order to obtain single crystal films. The optical loss coefficient of one such $BaTiO_3$ film was less than 0.5 dB/cm. It appears that when the large cations are in the first layer then epitaxial deposition of these perovskite ferroelectrics is destroyed.

Yeo et al.[25] were able to measure the optical loss due to a single grain boundary in bulk $Sr_{0.61}Ba_{0.39}Nb_2O_6$ and found a 1.1% scattering loss which suggests that the optical loss in polycrystalline films will be high. Kingston et al.[26] have found optical losses less than 5 dB/cm in $LiNbO_3$ in films consisting of columnar grains with roughness less than 1 nm and have succeeded in generating green light in such films by frequency-doubling. However, these columnar grains were similarly oriented grains having only small angle boundaries between them. Among the defects that can affect surface roughness are outgrowths on $LiNbO_3$ grown on Al_2O_3, which are believed due to imperfections in the substrate surface, and consist of grains that are oppositely polarized with respect to the remainder of the film.[26]

Ferroelectrics are not the only materials used for thin film wave guides. Organic based thin films, composite thin films consisting of very small SiO_2 and TiO_2 particles in a polymer base, SiO_2 films and various ion-implanted films have produced usable wave guides with low optical losses. Naturally, the relations between structure and properties vary between these materials. The implantations and incorporations of inorganic particles serve the purpose either of providing ions that luminesce and produce optical gain or of increasing the refractive index of the host material. These hosts are amorphous and will usually be homogeneous. The only structural elements in the composite material that can produce optical losses are inhomogeneities, such as voids, which under normal film deposition procedures are absent, or too large particles, which can be precipitated or induced to precipitate by charge–transfer complexes. In the glass hosts light-absorbing defects produced in the implantation process, such as dangling bonds, or scattering centers, such as displaced atoms, can contribute to optical loss and the dose and subsequent annealing needs to be optimized to eliminate such defects. Porous silicon can also be produced on silicon surfaces in defined areas, as, for example, by anodization, and act as wave guides in integrated silicon applications. The refractive index of the porous silicon is controllable over a wide range by manipulation of its density.[67]

3. Passive optical properties.

3.1. Refractive index.

The refractive index of a film is a monotonic function of the density. Thus, the removal of void networks at columnar boundaries, as can be achieved by appropriate energetic particle bombardment during deposition or by deposition above the transition temperature T_1 (see I), is the major effect of film structure on refractive index. Indeed, in the past, values of the index of refraction of the thin film and of the bulk material were used to determine the so-called packing fraction, which provides a measure of the film density compared to the ideal density of the bulk material. Currently, with the deposition techniques available, such as ion-assisted deposition, thin films for optical coatings are produced with near ideal densities. The packing fractions that have been evaluated in Zone 1 deposition using measured values of the refractive index range from a low of 0.57 on CaF_2, with the median falling between 0.7 and 0.8. Thus, the routine attainment of a packing fraction of unity at this time represents a considerable advance in the art of depositing optical films. (It is ironic, considering the extreme degree of specialization in research practiced at the present writing, that the void network structure characteristic of zone 1 was rediscovered by different groups of investigators as a consequence of the vital effect of this structure on those film properties uniquely significant to each group, e.g. dangling bond density in amorphous silicon, refractive index in optical films, noise in hard magnetic films.)

Crystallographic structure also has an effect on refractive index in that for non-cubic structures the refractive index is a maximum in the direction of highest linear density, i.e. the direction of closest atom packing. Hence, for non-cubic polycrystalline films the refractive index should be sensitive to texture.

3.2. Solar absorptance.

The property which measures the ability of a film to convert incoming radiation into heat is the solar absorptance and this property is sensitive to the surface morphology and to the microstructure of the film. For example, a dendritic metal surface with interdendrite dimensions on the order of the wavelength of the solar spectrum, which allows the radiation to penetrate into the interdendritic space, but not to exit from it, provides excellent values of the solar absorptance. Also, a film that consists of a distribution of metal particles in a transparent medium, with particle and interparticle dimensions within the emission wavelengths of the solar spectrum, functions to absorb the solar radiation in the same efficient manner.

A recently discovered effect of "crystal structure" on solar absorptance is provided by the observation that quasi-crystalline films of Al–Cu–Fe exhibit nearly ideal properties for solar absorber applications in that they have high absorptance at short wavelengths and low absorptance at wavelengths above 2 nm.[59] This peculiar combination of properties is determined obviously by the quasi-crystalline nature of the material in these films.

3.3. Absorption coefficient.

Absorption in the visible spectrum can be introduced into an inherently transparent material through microstructural manipulation. In this case, the introduction of pores containing internal surfaces which scatter light corresponds to the microstructural change that transforms a transparent medium to a light absorbing one. Just the reverse microstructural rearrangement, the elimination of pores, was responsible for converting the normally white, non-transparent alumina into a transparent substance, a discovery that led to commercial rewards. In thin films, the equivalent of the pores is the void network and/or lower density regions between grains. Consequently, thin dielectric films produced by evaporation usually have absorption coefficients that are an order of magnitude larger (usually $\sim 10 \, \text{cm}^{-1}$) than obtained in bulk dielectric materials ($\sim 1 \, \text{cm}^{-1}$). However, the use of energetic ion beams should eliminate the low-density regions and yield films with properties close to bulk values. Some evidence has been presented which indicates that films of Indium-Tin-oxide deposited onto an amorphous substrate while under energetic bombardment of the surface and having grain boundaries exhibited the same electrical properties, and presumably the same transparency, as a monocrystalline film of ITO deposited under the same conditions, but with an epitaxial substrate in place of an amorphous one. Electron-beam vapor deposited films exhibit higher electrical resistivity and presumably higher absorption coefficient.

It is also possible under improper deposition conditions to introduce impurities into the films being deposited. If these impurities correspond to color centers in the dielectric then the absorption coefficient will increase over that for the pure material.

There are many more passive optical properties that have not been considered in this section. However, the relationships between structure and these properties are limited. The reader is referred to numerous texts on the subject of passive optical properties for a listing of these properties. (e.g. see Chapter 11 in *The Materials Science of Thin Films*, M. Ohring, Academic Press, 1992.) However, the effects of structure, other than thickness and surface roughness, on these properties are negligible as already noted and, thus, a discussion of them is not warranted in this book.[59]

4. Active optical properties.

4.1. Thermochromism.

Thermochromism is the phenomenon in which a change in temperature induces a change in light transmittance or absorption of a material. One may expect such marked changes in the optical properties when a material undergoes a transition from a semiconductor phase to a metallic one. There are several materials that exhibit semiconductor to metallic transitions upon increase of temperature. For example, VO_2 transforms from a semiconducting monoclinic structure to a metallic tetragonal structure at about 67°C. Because this transformation is diffusionless the switching time for it is very short. Values less than 10 psec have been reported.[30] Because there is a shape change associated with the transformation and because films are constrained by the substrate, stresses will be introduced upon a phase transformation. The effect of such stresses in repeated cycling of the phase transition is not known, but such repeated cycling should lead to deterioration of the film. The main potential applications of this phenomenon are in optical switching and optical limiting, as well as optical recording. However, the use of diffusionless phase change materials for such applications is limited at the present writing by the cycle lifetime prior to degradation.

One of the more interesting, and apparently commercial, examples of thermochromism involves the difference in absorption coefficient and reflectivity between crystalline and amorphous structures of a material. Erasable phase-change optical recording media make use of this effect. The crystalline phase consisting of fine grains has a higher absorption coefficient, and higher reflectivity, than the amorphous phase. Usable media must have an absorption edge that shifts in the visible or near-infrared wavelength region with phase transition; a suitable melting point above that for self-crystallization and one which an applicable laser can exceed in a pulse; and a phase transition that is rapid and stable. These criteria have been satisfied by compositions along the pseudo-binary line between the compound GeTe and Sb_2Te_3.[31] Writing is accomplished by a pulse that rapidly melts a crystalline matrix

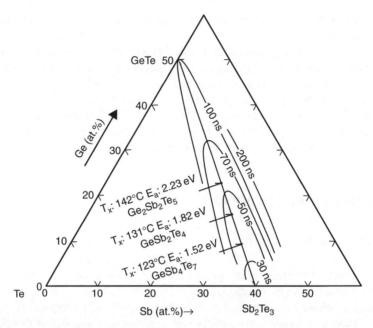

Figure 3.3. Composition dependence of the minimum laser-irradiation durations to cause crystallization in 100 nm thick films sandwiched between 100 and 200-nm-thick ZnS layers. The laser power is fixed at 8 mW on the samples. (After N. Yamada, MRS Bull. <u>21</u>, 48(1996) with permission.)

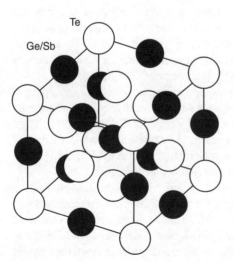

Figure 3.4. NaCl structure with Te atoms on points of one face centered cubic (fcc) lattice and Ge and Sb sharing the points of the other fcc lattice for the compounds along the pseudobinary line between the GeTe and Sb_2Te_3 compounds.

to produce an amorphous spot on rapid freezing. Erasing is accomplished by a less powerful laser pulse that crystallizes the amorphous spot. The speed of the latter transition was the factor limiting application of this method of optical recording. A fascinating application of materials science was involved in the search for a material that would reduce the erase time from several microseconds to one that was commercially viable (30 ns).[32] Normally, the low temperature required for melting by a laser pulse limited the use of such phase change materials to eutectic based systems. However, eutectics involve the solidification of two different solid phases and, hence, appreciable diffusion lengths to achieve the required separation of components. On the other hand, solidification of a compound composition does not involve such long range diffusion of components. Thus, candidate phase change materials were limited to compound compositions. In these materials the erase time is controlled by the nucleation and growth of crystals having the same composition as the metastable amorphous host. The activation energies involved in the nucleation and growth processes will be smaller the lower are the binding energies between the components of the compound material and the lower are the melting points of the components. These considerations led to the choice of a system containing Te, in place of one containing Se. Since the Te–M binding energies are less than the Se–M binding energies (i.e. the electronegativity of Se exceeds that of Te, both of which exceed those of Ge and Sb). These concepts bore fruit as shown in Figure 3.3 by the trends in the measured activation energy for, and time of, crystallization with increasing concentrations of the low-melting point components in the compound. The line between GeTe and Sb_2Te_3 in the figure represents a pseudobinary compound system having the same crystal structure, the NaCl structure, where Te occupies the points of one face-centered lattice and the Ge and Sb atoms occupy the points of the other face-centered lattice, as shown in Figure 3.4. Apparently, the cyclability and use lifetime of this product, that depend on diffusional transitions, are sufficiently better than those for diffusionless ones for it to be commercial.

4.2. Electro-optical properties.

4.2.1. Polarization and electro-optic coefficients.

Non-linear optical phenomena include second-harmonic generation, sum- and difference-frequency generation, modulation, switching, etc. The non-linear optical properties that give rise to these phenomena involve either the dependence of the polarization of the charges in the optical material on the electric field or a change in the absorption coefficient due either to the incident light or an electric field.

The dependence of the polarization P upon the applied field E may be expressed in the form

$$P = \varepsilon_0[\chi^{(1)}E + \chi^{(2)}E^2 + \chi^{(3)}E^3 \dots] \qquad (2.1)$$

where ε_0 is the permittivity of free space, $\chi^{(1)}$ is the linear susceptibility, and the quadratic and third-order susceptibilities are $\chi^{(2)}$ and $\chi^{(3)}$. These susceptibilities display the symmetry properties of the optical material. The coefficients of the associated tensors are r_{ij}, d_{ij}, and g_{ijk}, respectively.

There are three main classes of electro-optic materials: ferroelectrics, semiconductors, and organics. In terms of the change in refractive index per applied electric field ferroelectric materials are more sensitive than organics and semiconductors. However, when the application requires low dielectric constant as well, then the organics are the superior material class. The linear electro-optic coefficients are non-zero in materials that do not have inversion symmetry. Such materials have two orthogonal optic axes. Associated with these axes are two different indexes of refraction- an ordinary index n_o and an extraordinary index, n_e. An electric field applied parallel to one of the optic axes changes one refractive index with respect to the other, thus changing the birefringence and the state of polarization of light propagating through the material.

4.2.2. Linear electro-optic properties.

The property that is significant with respect to light beam modulation or switching applications is the change in refractive index of the material induced by an applied electric field. This effect depends on the product $n_i^3 \cdot r_{ij}$ through the relation $\Delta n = 0.5\, n_i^3 \cdot r_{ij} E_j$, where n_i is the refractive index, r_{ij} is the linear electro-optic tensor coefficient, and E_j is the applied field. As shown in Table 3.1, ferroelectrics have the highest values of $n_i^3 \cdot r_{ij}$. Thus, there has been a concentrated effort to produce ferroelectric thin films.

Now, for second-harmonic-generation in ferroelectric thin film wave guides in a frequency doubling device relying on copropagating modes, optical losses of 2 dB/cm reduce the conversion efficiency by about half. Since we have noted in Section 2 of this chapter that a single grain boundary produces a scattering loss of 1.1%, it is obvious that whatever scheme is used for SHG it cannot be achieved in a polycrystalline film. Thus, one aspect of structure for SHG is the requirement that the film involved be

Table 3.1. Values of $n^3 r$ for various electro-optic materials

EO materials	$n^3 r$ (pm/V)
Ferroelectrics	
$LiNbO_3$	320
$Sr_{1-x}Ba_xNb_2O_6$	2,460
$BaTiO_3$	11,300
KH_2PO_4	86
Semiconductors	
GaAs	43
InP	52
CdTe	152
Organics	
2-Methyl4-Nitroaniline	530
m-Nitroaniline	97

monocrystalline. In principle, SHG could be accomplished by periodic inversion of the domains and by proton exchange that periodically deadens the non-linear coefficient. The latter can be accomplished with a lesser chance of introducing defects that increase the optical loss, but at the cost of obtaining 25% of the conversion efficiency corresponding to the ideal domain inversion.

A change in the refractive index induced by an applied electric field results in a change in phase of the light beam traversing the region of altered refractive index. Use is made of this phenomenon in phase modulators, in intensity modulation of a light beam by control of the phase lag between two coherent beams to range from constructive to destructive interference, and in light switches. Since these applications involve the use of thin film wave guides the discussion in Section 2 of this chapter regarding the effect of structure on optical losses applies here as well.

4.2.3. Non-linear electro-optic properties.

4.2.3.1. Photorefractive effect.

Non-linear optical effects also arise from absorption in the optical media. The photorefractive effect occurs when photoexcited charge carriers separate and become trapped to produce a non-uniform space charge distribution that can modulate the refractive index via the electric-optic effect. If the separation of the charge carriers is induced by gradients in the illumination intensity, then it is possible by using two coherent light beams that form an interference pattern in the photorefractive medium to develop a periodic space charge. Thus, a photorefractive material can be used to produce a hologram. Both organic and inorganic candidate photorefractive materials are at this writing being actively investigated. One of the materials that are candidates for photorefractive application is a composite material consisting of a photorefractive host, such as vitreous silica, $LiNbO_3$, MgO, and Al_2O_3, which is embedded with nanocrystals of various metals introduced by ion implantation and precipitation. A third order non-linear refractive index of $5 \cdot 10^{-8}$ esu was reported for one such composite film.[69] The high third order non-linear susceptibility of these materials results from the dramatic enhancement of the local optical field in the vicinity of the metal nanoparticles at the wavelength of the surface plasmon resonance.[70] Thus, structure affects all the prerequisites for the photorefractive effect: generation, mobility, and trapping of charge carriers and the magnitude of the electro-optic coefficients.

4.2.3.2. Electrochromism.

Electrochromism of materials is a phenomenon in which an electric field induces a change in either the light reflectance spectrum or the light absorption spectrum of the material. Application of the electric field induces either the intercalation or the removal of positive ions and electrons and thereby achieves the change of state associated with the change in light reflectance or absorption properties. Such intercalation and removal of charge carriers requires efficient coupling of the electrochromic material to electrolytes, preferably solid ones, that supply and absorb the charge carriers. This can be best accomplished when the electrochromic material has a large surface area per unit volume. The property affected by this structural parameter is the speed of switching on and off of the electrochromic response. In particular, it has been found[43] that the density of WO_3 prepared by electrodeposition was about 4.8–5.0 g/cm^3, while that for vacuum evaporated films of this material was found to be 5.3–5.7 by one group[44] and 6.5 g/cm^3 by another one.[45] The switching time for the electrodeposited film was found to be less than or equal to 0.1 secods, while for the vacuum deposited films this time is about two orders of magnitude larger. Also, for NiO films, it has been found that the switching speed depends upon structure in some as yet unknown way in that films produced by oxidation of co-evaporated Ni–C films had a switching time of 1 second compared to one of 8 seconds for films produced by sputtering and electrodeposition.[46] An analysis of results suggested that the electrochromic reaction occurred at the grain boundaries in the former films, but was homogeneous in the latter films. This may be a consequence of the existence of a grain boundary void network in the former films which is absent in the latter. The switching speed is important for potential display applications, but is not significant for "smart" window applications involving control over heat input and output. Also, switching time is not significant for an electronic book, that would use an electrochromic material with a sufficient associated change in the absorption coefficient to produce a contrast in reflected, rather than transmitted, light in order to reduce eye strain.

Since the electrochromism phenomenon involves intercalation of positive ions, such as Li^+ or H^+ into the electrochromic material it is possible that the structure of the latter may be affected by the intercalation of such ions, which must act to change the specific volume as well as the stress in the intercalated film adhering to a substrate. It is believed that this process occurs in viologens, which when deposited are amorphous, but which with cycling of the applied voltage undergo transformation to the crystalline state with change in the electrochromic properties.[47]

The crystalline state of the electrochromic material WO_3 also affects its electrochromic properties. Uncolored WO_3 has a lower transmittance in the amorphous state than in the crystalline state in the infra-red spectrum for wavelengths shorter than about 3.5 μm and vice versa for longer wavelengths.[48] Since the diffusivity of the positive intercalating ions is not likely to be the same in an amorphous host as compared to a crystalline host one may expect the switching time to display this difference. Further, because the electronic band structures of these two different phases will be different one may expect that the details of the absorption spectrum for the colored phases will also differ.

4.3. Non-linear electro-optic materials.

At this writing it is not possible to predict or interpret macroscopic non-linear optical properties from microscopic models for any of the electro-optic materials. We limit the following discussion therefore to a description of what is known phenomenologically.

4.3.1. Ferroelectric thin films.

Domains in monocrystalline film may not be aligned, as also occurs in bulk materials. Thus, Wessels et al.[24] found that poling* increased the effective second-order non-linear electro-optic (EO) coefficient, d_{eff}, from 6 to 10 pm/V for $KNbO_3$ on a $MgAl_2O_4$ substrate. However, the $KNbO_3$ monocrystalline films grown on $LaAlO_3$ substrates were characterized by single domains and had an effective second-order non-linear EO coefficient of $d_{eff} = 13$ pm/V.[24] (Given that X-ray diffraction may not detect these domains it is understandable that films having different domains are still denoted monocrystalline even if they are not strictly monocrystalline.) Thus, it appears that in $KNbO_3$, at least, the production of a single domain in a monocrystalline film maximizes the effective second order non-linear electro-optic coefficient.

$KNbO_3$ "monocrystalline" films grown on MgO had a value of the d_{31} coefficient equal to 5.1 pm/V, whereas "monocrystalline" films grown on (MgO) $(Al_2O_3)_{2.5}$ spinel had values of this coefficient ranging from 2 to 4.5 pm/V.[23] Bulk single crystals of this material have $d_{31} = 15$ pm/V and $d_{33} = 27$ pm/V. Since these substrates have different lattice parameters and thermal expansion coefficients we may expect that the stresses in the various $KNbO_3$ films on these substrates differ. From the discussion on the effect of stress on domain orientation in Chapter I, Section 5.1, we may conclude that these films also differ in their domain configurations. Given the effect of domain configuration on d_{eff} discussed in the previous paragraph, we may further conclude that the difference in d_{31} among these films is due to a difference in the domain configuration.

Although the presence of two in-plane crystal orientations does not appear to bring about a difference between the optimum EO coefficients observed for (001) textured thin films of $Sr_xBa_{1-x}NbO_3$ on (100) MgO substrates compared to values of these coefficients for bulk single crystals, the measured second order d_{eff} coefficient varies with the microstructural perfection as revealed by the X-ray rocking curve FWHM values, i.e. the smaller the FWHM value the larger the d_{eff} value.[33] However, it must be noted that the value of X, the composition of Sr relative to Ba, also varied monotonically with the FWHM. Thus, it is possible that the EO coefficient depends upon the X value rather than on the FWHM values.

---

* Poling is the process of subjecting the ferroelectric material to an electric field in order to increase domain alignment.

Orientation of $LiNbO_3$ films affects the quadratic EO coefficient. For example, (0001) films have negligible quadratic EO coefficients while (11 $\bar{2}$ 0) films have appreciable values of the quadratic EO coefficient, e.g. an effective value of $2.38 \cdot 10^{-15} m^2/V^2$.[34]

4.3.2. Organic thin films.

Organic materials achieve their non-linear optical properties through the incorporation of dipole molecules and the alignment of these molecules to produce the polarization. There are very many additional requirements for a satisfactory electro-optical organic material, but we will not be concerned with these requirements in this section. The primary effect of structure on the non-linear optical properties of organic thin films is likely therefore to be a consequence of an effect on the spectrum of dipole orientations.

In one class of organic thin films, the host is polymeric and the chromophores (dipole molecules) may or may not be bonded to the polymers. Alignment of these dipole molecules is usually achieved by cooling under an applied electric field to below the glass temperature or in the case of thermosetting polymer hosts by heating in the presence of an electric field above the bonding temperature. Such films would be expected to have a rather broad spectrum of dipole orientations and a respectable value of the second order EO coefficient, d_{33}, of 60 pm/V has been reported[37] from films produced via these methods.

Another method of orienting the dipole molecules involves placing a polymer solution containing individual non-centro symmetric polymers under an electric field during evaporation of the solvent.[38] Still another makes use of laser assisted electric field poling to orient azobenzene chromophores in the plane of the poling field[39] prior to freezing-in the orientation via one of several methods. Others involve vapor deposition in the presence of an electric field,[40] molecular self-assembly methods,[41] etc. At this writing the author is not aware of any studies relating the spectrum of dipole orientations within a domain to the non-linear optical properties other than the one reference[39] cited above. In this case, removal of randomness in the orientation of the dipole molecules about the poling axis, as induced by the use of plane-polarized laser illumination during poling, results in marked change in the ratio of the electro-optics coefficients, i.e. $r_{33}/r_{13} = 7$, a value much greater than that found in the absence of the laser illumination with similar results for the second harmonic coefficient anisotropies, d_{33}/d_{13}.

4.3.3. Semiconductor thin films.

Appropriately grown semiconductor structures which yield a separation of oppositely charged species also have an associated electric polarization and can yield non-linear electro-optic properties. The main role of structure in this case is that associated with the morphology of the quantum wells required to contain the electric charges. Separation of oppositely charge carriers in a quantum well layer by an applied field allows the quantum well to modulate transmitted light. This effect is known as the quantum confined Stark effect (QCSE). GaAs/AlGaAs and InGaAs/ AlGaAs superlattices, where the layer thickness for the quantum well layer is less than 100 Å, are examples of such semiconductor elecro-optic modulators. In another example, the use of sufficiently thin GaAs quantum well layers alternating with AlAs layers allows a separation of holes and electrons, such that the holes occupy the GaAs layers and the electrons the AlAs layers. In principle, larger electro-optic coefficients should be obtainable in quantum dots. Quantum wires have been produced by spinodal decomposition induced by internal stress in GaInAs and GaInP with unique lasing properties.[42] The A, B, m, n and substrate system parameters must be controlled in the deposition of $(III_A-V)_m$ and $(III_B-V)_n$ superlattices to be able to produce the quantum wires during deposition.

Recapitulation

The band gap energy of semiconductors having a given crystal structure is a function of the lattice parameter. The latter can be changed by strain induced by epitaxy and by long range ordering of constituents. Non-radiative recombination of electrons and holes acts to reduce the luminescence efficiency by reducing the number of radiative recombination events. Non-radiative recombination occurs at structural defects, such as dislocations, point defects, and interfaces and the release in energy upon recombination can induce the generation of additional point defects.

Er^{3+} in Si can act as a source of luminescence and its efficiency in this regard can be enhanced by complexing with O and F atoms. Much greater increase in the efficiency at room temperature can be obtained from Er^{3+} ions located in carrier free regions (depletion region, insulating region of avalanche diode) using hot electrons in reverse bias under diode breakdown condition. Porous silicon also acts to enhance luminescence from silicon either by providing regions that act to quantum confine charge carriers or by providing a large surface area per unit volume of silicon along which O and H atoms can adsorb. The intensity of illumination from porous silicon increases with increase in the surface area per unit volume. Surface non-radiative recombination sites must be passivated, however, for the luminescence to be efficient. Luminescence can be induced in quantum wells. For SiGe/Si-strained multilayer quantum wells to luminesce efficiently, the Si layers must be free of Ge atoms and the population of non-radiative recombination centers must be nil. This is achieved by MBE at high substrate temperature in the presence of a surfactant layer. Finally, organic thin films of various types can host luminescence, which is sensitive to the structure of the host polymer and to the dopants added to the film.

Absorption losses in thin film wave guides are produced by grain boundaries, vacancies, and impurities. Scattering losses are produced by surface roughness, and material inhomogeneities, such as second phases, twins, grain and domain boundaries and voids. In amorphous hosts dangling bonds or displaced atoms can contribute to optical loss.

Refractive index of thin films differs from that of bulk material only in films deposited with the substrate temperature in zone 1 because this property depends uniquely on the density for a given material. The intergranular void network produced in thin films deposited at such substrate temperatures is responsible for these films having lower density than their bulk counterparts.

Phase transitions producing changes in the absorption coefficient represent the structural effect in the phenomenon of thermochromism. Also, impurities that act as color centers in dielectric materials obviously affect the absorption coefficient.

Any structural discontinuities, such as grain boundaries, surface roughness, and perhaps all lattice defects from point- to extended defects that contribute to optical losses affect properties dependent on the linear electro-optic coefficients.

The structural aspects that appear to affect the non-linear optical properties of thin ferroelectric films are: texture, mono- versus polycrystallinity and film stress. The effect of structure on the electro-optic properties of organic thin films consists of the effect of the spectrum of orientations of dipole molecules existing in these films. In semiconductors, structure in the form of quantum wells allows for the development non-linear optical properties by producing and/or confining the separation of holes and electrons along defined orientations.

Defect structure can be vital to the functioning of the photorefractive effect by providing the means of trapping separated photoexcited charge carriers of opposite sign.

The phenomenon of electrochromism is sensitive to those aspects of structure that affect the transport velocity of intercalated positive ions.

References

1. A. Gomyo, T. Suzuki, K. Kobayashi, S. Kawata and I. Hino, Appl. Phys. Lett. 50, 673(1987).
2. T. Egawa, Y. Hasegawa, T. Jimbo and M. Umeno, Appl. Phys. Lett. 67, 2995(1995).
3. R.L. Gunshor and A.V. Nurmikko, MRS Bull. XX, 15(1995).
4. O. Ueda, T. Fujii and Y. Nakata, Proc. Int. Conf. Sci. and Tech. of Defect Control in Semiconductors, Yokohama, Japan, Elsevier, Amsterdam, 1989.
5. H. Takasugi, M. Kawabe and Y. Bando, Jpn. J. Appl Phys. 26, L584(1987).
6. (a) O. Ueda, S. Isozumi and S. Komiya, Jpn. J. Appl. Phys. 23, L241(1984); (b) S.N.G. Chu and S. Nakahara, MRS Symp. Proc. 421, 407(1996).
7. S.D. Lester, F.A. Ponce, M.G. Crawford and D.A. Steigerwals, Appl. Phys. Lett. 66, 1249(1995).

8. O. Ueda, MRS Symp. Proc. 262, 197(1992).
9. W.M. Chen, I.A. Buyanova, A. Henry, W.-X. Ni, G.V. Hansson and B. Monemar, MRS Symp. Proc. 378, 135(1995).
10. S.M. Lord, G. Roos, B. Pezeshki and J.S. Harris, Jr., MRS Symp. Proc. 262, 881(1992).
11. F.Y.G. Ren, J. Michel, Q. Sun-Paduano, B. Zheng, H. Kitawa, D.C. Jacobson, J.M. Poate and L.C. Kimerling, MRS Symp. Proc. 301, 87(1993).
12. J. Michel, J.L. Benton, R.F. Ferrante, D.C. Jacobson, D.J. Eaglesham, E.A. Fitzgerald, t.-H. Xie, J.M. Poate and L.C. Kimerling, J. Appl. Phys. 70, 2672(1991).
13. J.J. Pradissitto, M. Federighi, C.W. Pitt, W.P. Gillin, A.G. James and R.J. Wilson, MRS Symp. Proc. 392, 217(1995).
14. T. Taskin, Q. Huda, A. Scholes, J.H. Evans, A.R. Peaker, P. Hemment, C. Jeynes and Z. Jafri, MRS Symp. Proc. 392, 223(1995).
15. (a) J. Palm and L.C. Kimerling, MRS Symp. Proc. 378, 703(1995); (b) F.Y.G. Ren, J. Michel, L.C. Kimerling, D.C. Jacobson, Y.H. Xie, D.C. Eaglesham, E.A. Fitzgerald and J.M. Poate, J. Appl. Phys. 70, 2667(1991); (c) J.V. Gates, A.J. Bruce, J. Shmulovich, Y.H. Wong, G. Nykolak, M.R.X. Barros and R. Ghosh, MRS Symp. Proc. 392, 209(1995).
16. (a) L.T. Canham, Appl. Phys. Lett. 57, 1046(1990); (b) V. Lehmann and U. Gösele, Appl. Phys. Lett. 60, 856(1991); (c) A.G. Cullis and L.T. Canham, Nature 353, 335(1991); F. Koch et al., Proc. 20th Int. Conf. Semiconductor Physics, World Scientific, Singapore, 1992, p. 148.
17. (a) C. Pickering, in Porous Silicon, eds. Z.C. Feng and R. Tsu, World Scientific, Singapore, 1994, p. 3; (b) M. Stutzmann and M.S. Brandt in ibid., p. 417; (c) Y. Kanemitsu, T. Matsumoto, T. Fugati and H. Mimura in ibid., p. 363.
18. (a) S.Z. Weisz, J. Avalos, M. Gomez, A. Many, Y. Goldstein and E. Savir, MRS Symp. Proc. 378, 899(1995); (b) K. Li, D.C. Diaz, J.C. Campbell and C. Tsai in Porous Silicon, see Ref. 19, p. 261; (c) P.M. Fauchet in ibid., p. 449; (d) Y. Kanemitsu, T. Ogawa, K. Shiraishi and K. Takeda, Phys. Rev. B48, 4883(1993).
19. (a) See articles in **Optical and Photonic Applications of Electroactive and Conducting Polymers**, S.C. Yang and P. Chandrasekhar, SPIE 2528 (1995);.also see articles in MRS Symp. Proc. 413 (1996), Part I; (b) R. Friend, ibid., also Q. Pei and Y. Yang, J. Am. Chem. Soc. 81, 3294(1997); Y. Cao, G. Yu, A.J. Heeger and C.Y. Yang, Appl. Phys. Lett. 68, 3218(1996).
20. (a) R. Gutman, J. Huliger, R. Hauert and E.M. Moser, J. Appl. Phys. 70, 2648(1991); (b) J.C. Brice, O.F. Hill, P.A.C. Whiffin and J.A. Wilkinson, J. Crys. Growth 10, 133(1971).
21. D.K. Fork, F. Armani-Leplingard and J.J. Kingston, Proc. of the Fall 1994 Symposium on Ferroelectric Thin Films II.
22. J.J. Kingston, F. Armani-Leplingaard, D.K. Fork and G.B. Anderson, MRS Symp. Proc. 401, 243(1006).
23. S. Schwyn-Thöny, MRS Symp. Proc. 341, 253(1994).
24. B.W. Wessels, M.J. Nystrom, J. Chen, D. Studebaker and T.J. Marks, MRS Symp. Proc. 401, 211(1006).
25. J.S. Yeo, K.E. Youden, T.F. Huang, L. Hesselink and J.S. Harris, Jr., MRS Symp. Proc. 401, 225(1996).
26. J.J. Kingston, D.K. Fork, F. Leplingard and F.A. Ponce, MRS Symp. Proc. 341, 289(1994).

27. C.M. Foster, Z. Li, G.R. Bai, H. Fou, D. Guo and H.L.M. Chang, MRS Symp. Proc. 341, 295(1994).
28. R.A. McKee, F.J. Walker, E.D. Specht and K.B. Alexander, MRS Symp. Proc. 341, 309(1994).
29. Y. Shigesato, I. Yasui and D.C. Paine, J. Met. 47, 47(1995).
30. M.F. Becker et al., Springer Series in Chemical Physics, 4, Picosecond Phenomena, 1978, p. 236; A. Zylverstejn et al., Phys. Lett. 54A, 145(1975).
31. N. Yamada, E. Ohno, K. Nishiuchi, N. Akahira and M. Takao, J. Appl. Phys. 69, 2849(1991); R. Imanaka, T. Saimi, Y. Okazaki and I. Kawamura in Proc. Int. Symp. Optical Memory 1995, Technical Digest 41(1995).
32. N. Yamada, MRS Bull. 21, 48(1996).
33. M.J. Nystrom, B.W. Wessels, D.A. Neumayer, T.J. Marks, W.P. Lin and G.K. Wong, MRS Symp. Proc. 361, 167(1995).
34. S.-H. Lee, T.W. Noh and J.-H. Lee, MRS Symp. Proc. 401, 261(1996).
35. H. El Ghitani and S. Martinuzzi, Mat. Sci. Eng., B4, 153(1989).
36. A.K.-Y. Jen, T.-A. Chen, V.P. Rao, Y.-M. Cai, Y.-J. Liu, K.H. Drost, R.M. Mininni, L.R. Dalton, P. Bedworth and S.T. Marder, MRS Symp. Proc. 392, 33(1995); M.N. Mang et al., ibid., p. 62; L. Li et al., MRS Symp. Proc. 277, 161(1992).
37. X. Yang, D. McBranch, B. Swanson and DeQ. Li, MRS Symp. Proc. 392, 27(1995).
38. D.J. Trantolo, J.D. Gresser, D.L. Wise, M.G. Mogul, T.M. Cooper and G.E. Wnek, in **Optical and Photonic Applications of Electroactive and Conducting Polymers**, eds. S.C. Yang and P. Chandrasekhar, Proc. SPIE 2528, 219(1995).
39. L.R. Dalton, A.W. Harper, J. Zhu, W.H. Steier, R. Salovey, J. Wu and U. Efron, in ibid., p. 106.
40. T. Yoshimura, S. Tatsura and W. Sotoyama, Thin Solid Film. 207, 9(1992).
41. S. Yitzchaik, S.B. Roscoe, A.K. Kakkar, D.S. Allan, T.J. Marks, Z.Y. Xu, T.G. Yang, W.P. Lin and G.K. Wong, J. Phys. Chem., 97, 6958(1993).
42. K.Y. Cheng and K.C. Hsieh, MRS Symp. Proc. 417, 241(1996).
43. P.K. Shen, K.Y. Chen and A.C.C. Tseung, Proc. Symp. Electrochromic Materials, 94-2, The Electrochemical Society, Pennington, NJ, 1994, p. 14.
44. N. Yoskiike, M. Ayusawa and S. Kondo, J. Electrochem. Soc. 131, 2600(1984).
45. S.K. Deb, Phil. Mag. 27, 801(1973).
46. Y. Sato, S. Tamura and K. Murai, in **Electrochromic Materials and Their Applications III**, eds. C.R. Greenberg and K.-C. Ho, PV 96-24, The Electrochemical Society Proceedings Series, Pennington, NJ, 1997.
47. N.J. Goddard, A.C. Jackson and M.G. Thomas, J. Electroanal. Chem. 159, 325(1983).
48. K.A. Gesheva, G. Stoyanov and D. Gogova, MRS Symp. Proc. 415, 155(1996).
49. Z. Jing, G. Lucovsky and J.L Whitten, MRS Symp. Proc. 318, 287(1994) and references cited in this article.
50. S. Fukatsu, N. Usami and Y. Shiraki, Jpn. J. Appl. Phys. 32, 1502(1993); S. Fukatsu, N. Usami, T. Chinzei, Y. Shiraki, A. Nishida and K. Nakawaga, Jpn. J. Appl. Phys. 31, L1015(1992); Y. Kato, S. Fukatsu, N. Usami and Y. Shiraki, Appl. Phys. Lett. 63, 2414(1993).
51. A. Thilderkvist, J. Michel, S.-T. Ngiam, L.C. Kimerling and K.D. Kolenbrander, MRS Symp. Proc. 405, 265(1996).
52. A.A. Seraphin, S.-T. Ngiam and K.D. Kolenbrander, ibid., p. 277.

53. S. Veprek, T. Wirschem, M. Rückschloβ, C. Ossaknik, J. Dian, S. Perna and I. Gregora, ibid., p. 141.
54. B. Chatterjee and S.A. Ringel, J. Appl. Phys. 77, 3885(1995).
55. It is interesting to note that a similar phenomenon is known to occur in the Staebler–Wronski effect that is found a-Si subject to light illumination, which we will discuss in detail in Chapter VII.
56. J.C. Phillips, **Bonds and Bands in Semiconductors**, Academic Press, 1973, p. 242.
57. This concept is hinted in Abstract H5.4 for the MRS Fall 1997 meeting.
58. S.D. Hersee, J.C. Ramer and K.J. Malloy, MRS Bull. 22, 45(1997).
59. M.F. Besser and T. Eisenhammer, MRS Bull. 22(11), 62(1997).
60. J.M. Langer, A Suchocki, L.V. Hong, P. Ciepielewski and W. Walukiewicz, Physica 117B, 118B, 152(1983).
61. G. Franzo, F. Priolo, S. Coffa, A. Polman and A. Carnera, Appl. Phys. Lett. 64, 2235(1994); G. Franzo, S. Coffa, F. Priolo and C. Spinella, J. Appl. Phys. 81, 2784(1997).
62. L. Tsyveskov, K.D. Hirschman, P.M. Fauchet and V. Bondarenko, MRS Symp. Proc. 486, (1998).
63. M.L. Brongersma, K.S. Min, E. Boer, H.A. Atwater and A. Polman, ibid.
64. W. Wu, X.-F. Huang, K.J. Chen, J.B. Xu and D. Chen, ibid.
65. H.W.H. Lee, G.R. Delgado, C.H.M. Maree, ibid.
66. R. Pairleitner, C. Hochfilzer, S. Tasch, G. Leising, U. Scherf and K. Mullen, MRS Symp. Proc. 488, (1998).
67. M. Araki, H. Koyama and N. Koshida, Appl. Phys. Lett. 68, 2999(1996).
68. W.-X. Ni, C.-X. Du, K.B. Joelsson, G. Pozina, I.A. Buyanova, W.M. Chen and G.V. Hansson, MRS Symp. Proc. 486, (1998).
69. D. Ila, R.L. Zimmerman, E.K. Williams, C.C. Smith, S. Sarkisov, D.B. Poker and D.K. Hensley, MRS Symp. Proc. 504, (1998).
70. Another effect of structure on the photorefractive effect is exhibited in films that exhibit persistent photoconductivity as a consequence of a distribution of DX defects. See Chapter VII.
71. See, for example, **Concise Encyclopedia of Semiconducting Materials & Related Technologies**, eds. S. Mahajan and L.C. Kimerling, Pergamon Press, 1992, p. 263.

Appendix

A1. Significant effects of structure on optical properties of thin films reported from 1998 to 2005.

A1.1. Effect of defect structure on luminescence degradation.

The effective removal of dislocations from GaN by various techniques have been developed in this period. One such method is the epitaxial lateral over-growth ELO technique, which involves the use of a mask on the surface of a par-tially grown film. The mask has a hole which allows the epitaxial growth through the hole, without propagation of the threading dislocations nucleated at the sub-strate–film interface through the mask, and the lateral spread of the epitaxial film over the mask.[A1] Other techniques have been developed which similarly cutoff the propagation of the threading dislocations, at least in the vertical direction, in some cases bending the dislocations into a lateral direction.[A2] To the best of my knowl-edge no new defects have been found that induce luminescence degradation in addition to those already mentioned in this chapter.

The realization that the probability of non-radiative recombination of photogenerated charge carriers can be diminished by electrically isolating islands smaller than the diffusion length of photoluminescence material seems to have peaked in this period. Consequently, there have been a plenitude of studies of photoluminescence in nanoparticles isolated within transparent insulating hosts.

A1.2. Effect of defects and interfaces in enhancing luminescence.

Sensitizers have been found which increase the luminescent intensities achieved from Er in several hosts. In particular, Yb has been found to increase the PL intensity from Er in SiO_2 by a factor of 20 for visible as well as infra-red excitation.[A3]

A strange effect was found in magnetron sputtered a-Si:H films containing Er atoms, which were deposited under conditions of low H dilution so as to form very small crystalline units (<3 nm) within the amorphous matrix.[A4] This system exhibited a two order of magnitude increase in the Er-related PL intensity and a decrease of it by only a factor of five between 77 K and room temperature. Kik and Polman[A12] have provided an explanation for the latter result as well of a number of other observations. Kashkarov et al.[A5] found that the Si nanocrystals transferred energy to the surrounding Er ions with an increase in the 0.81 eV PL line intensity and a loss of the Si nanocrystal intrinsic PL band between 1.2 and 1.7 eV.

Experimental results implied that SiC bonds at the surface of Si nanocrystals embedded in a silica matrix provide for intense visible blue-white photoluminescence with incident 325 nm wavelength radiation from a HeCd laser.[A6] The smaller the Si nanocrystals and the higher the population density of these nanocrystals the higher should be the intensity of the photoluminescence. This feature was not investigated. If the implication is verified then these results suggest that interface states in this system (and perhaps other systems) can be centers of radiative recombination. This is certainly worth investigation inasmuch as interfaces are normally the sites of non-radiative recombination centers, e.g. dangling bonds. Similar results have been obtained with a SiN matrix populated with Si nanocrystals again with indications that the interface region is involved in the photoluminescence.[A7]

The possibility that the interface between nanoparticle and matrix is involved in optical gain has also been suggested.[A8] Indeed, optical gain has been observed and attributed to the recombination of strongly localized excitons at the Si-nc/SiO$_2$ interfaces.[A9] These interfaces are also sites of non-radiative recombination and hence for radiative recombination to occur freely the non-radiative sites must be passivated. This is generally accomplished by annealing in a hydrogen reactive atmosphere. The structure of the nanoparticles can vary from amorphous to crystalline depending upon the annealing history subsequent to deposition.[A10] Monodispersity of the Si nanoparticles contributes to the optical gain.[A11]

Electroluminescence in nanoparticles has been used to produce white light by control over the size of the particles. The particles in this case consisted of CdSe with a coating of ZnS.[A13] It was said that the coating improved the luminescence efficiency by saturating surface states. However, the luminescence stems from band to band excitation with the band gap and, hence, the wave-length controlled by the size of the particle. Thus, in this case the luminescence does not depend upon defects or interfaces except to the extent that non-radiative recombination is suppressed there.

Sites at dislocations in silicon electroluminesce and give rise to four peaks denoted by the letter D. However, other sites at dislocations are centers of non-radiative recombination. The latter are mainly impurities and an effort to remove these impurities via gettering and to passivate the intrinsic non-radiative recombination sites via hydrogen allowed the achievement at room temperature of a 0.1% external infrared electroluminescent efficiency from an array of dislocations having a density between 3 and 7 · 10^8 cm^{-2}.[A14] The observed electroluminescence stemmed from the D$_1$ peak. No band to band luminescence was observed in these experiments. Both the introduction of dislocations by implantation with B and the introduction of centers of radiative recombination, Er and β FeSi$_2$, were used to measure the effect of dislocations upon the luminescence produced in Si. The results were commendable in that strong silicon band edge luminescence was observed.[A18] The potential to produce a much higher population density of dislocations suggests that future attempts will be made to increase the electroluminescent yield from dislocated silicon.

A1.3. Organic and polymeric luminescent films.

Structure affects the luminescent efficiency of polymer LEDs. One type of structure is the macrostructure of the device and a manifestation of this structure is that of the hole- and electron-injection layers between the light-emitting polymer and the electrodes. Appropriate grading of these interlayers improves the luminescent efficiency.[A15] Not only the injection layers but also charge transfer blocking layers significantly affect the efficiency and output of the polymer LEDs. In particular, use of hole injection layers that consist of layer like compounds, which have all their surface bonds satisfied, coupled with oxide electron blocking layers act to move the recombination zone to the interface between the polymer LED and the electron blocking layer with a resultant high efficiency of luminescence.[A16]

A microstructural aspect that affects the luminescent efficiency of polymer LEDs, such as PPV, is that which determines the width of the distribution of exciton binding energies.[A17] The delocalization of the excitons leads to luminescent quenching as the field strength across the LED increases. The binding energy of the excitons increases with increasing dielectric constant which itself increases with increasing isotropy (structural disorder) of the polymer strands. Hence, spin casting is the preferable mode of forming the LED layer.

Structure in the form of photonic patterning of etch pits, optical gratings, photonic crystals designed to extract the light emitted inside the luminescent layer also acts to increase the light emitted to the outside.[A19]

The combination of CdSe quantum dots coated with ZnS acting as lumophores surrounded by polymer charge carrier films[A20] produced much better results than this quantum dot layer in the absence of the organic charge carrier films[A21]. The combination of phosphors of appropriate design with appropriate LEDs yield efficiency greater than that of incandescent and halogen lamps.[A22] Hence, the outlook is that white LED light sources will appear in the near future to compete with the sources previously available.

OLEDs are now commercial products and it is now known that with time OLEDs age producing deep traps by as yet an unknown process.[A23]

A1.4. Chromophores–thermo- and photo-.

Chromism refers to the change in optical absorption characteristics occasioned by a change in structure of the material. This change in structure may be induced by a change in temperature (thermochromism), exposure to light (photochromism) and a variety of other variables. The section on thermochromism in the original edition focussed on the phenomenon in inorganic materials. However, organic materials are also subject to thermochromism. This short review will focus on chromism in organic materials. A tutorial review of photochromism in crystalline organic materials has appeared recently.[A24] Consequently, it will not be repeated here except to point out that the applications of this effect in organic materials are in the areas of data storage, information processing, telecommunications, and optical switching. This is an active area of current research.

A1.5. Non-linear electro-optic materials.

A1.5.1. Ferroelectrics.

The significant advance in the application of ferroelectrics as non-linear electro-optic materials is the engineering of the domain pattern to fit the application. Control over the domain pattern via the use of patterned electrodes or ATM tip voltage application is the basis of the domain engineering concept. Periodic domain inversion allows for quasi-phase matching and hence nonlinear frequency conversion. Wavelength converters as well as continuously tunable lasers can be made. Such components are of great interest for, e.g., telecommunication, environmental sensing, and laser television systems. However, domain engineering has been limited mainly to bulk crystals rather than thin films. Only a few applications of domain engineering to thin films has been made. One of these involves liquid epitaxial grown films of $LiNbO_3$ for which the periodic poling was achieved using an electron beam.[A25]

A1.5.2. Organic thin films.

The design of organic electro-optic thin films has proceeded at a rapid pace to the point where such thin films have been included in devices. Modelling of the electro-optic properties has provided a basis for predicting the design of chromophore and polymer composite structure. One prediction that was borne out experimentally indicated that rotund shaped chromophores would yield larger electro-optic activity than chromophores with prolate ellipsoid shape. Another related to the supermolecular structure where in dendronized polymer structures covalent bonds prohibit "centrosymmetric" crystallization. A hypothetical ideal structure indicated by statistical mechanical based modelling is shown in Figure A3.1. This structure is non-centrosymmetric.[A26] Thus, not only does the molecular structure of the chromophore molecule affect the electro-optic properties, but also its shape and the stacking of the chromophores in the supermolecular unit affect the electro-optic properties as well. The trend in improvement of these properties is likely to continue in the near future as the knowledge gained from modelling studies enlarges the options for this improvement.

Among the devices produced with such electro-optic organic thin films are photonic radiofrequency phase shifter,[A27] time stretching devices for A/D conversion,[A28] ultra-high bandwidth oscillators,[A29] acoustic spectrum analyzers,[A30] optical

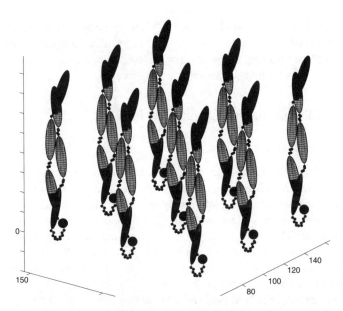

Figure A3.1. Predicted non centrosymmetric supermolecular structure consisting of chromophores coupled by flexible segments. Chromophores are the prolate ellipsoids with dipole moments as shown. The larger spheres represent a rigid assembly of several atoms that serve as the dendritic core of the structure. (Reproduced from L.R. Dalton, MRS Symp. Proc. 817, L7.2.1.(2004) with permission.)

switches and intensity modulators,[A31] laser beam steering devices and optical gyroscopes[A32] and wavelength selective filters for wavelength division multiplexing.[A26] For most of these applications the functioning depended upon as high a value of the electro-optic coefficient r_{33}. For the organic thin film materials developed in the near past the value of this coefficient and the other parameters exceed those for the best inorganic electro-optic materials. Indeed, there is little doubt that the thin film organic electro-optic materials will inspire a host of devices not yet conceived.

References to Appendix

A1.   A. Sakai, H. Sunakawa and A. Usui, Appl. Phys. Lett. 71, 2254(1997).
A2.   A. Sagar et al., Phys. Stat. Solidi (c) to be published 2005; S. Nata et al. MIJ-NSR, F99W2.8.
A3.   A. Kozanecki, K. Homewood, B.J. Sealy, Appl. Phys. Lett. 75, 793(1999).
A4.   M.V. Stepikhova et al., Phys. Sol. State 46, 113(2004).

A5. P.K. Kashkarov et al., Phys. Sol. State 46, 104(2004).
A6. S.-Y. Seo, K.-S. Cho and J.H. Shin, Appl. Phys. Lett. 84, 717(2004).
A7. Y.Q. Wang, Y.G. Wang, L. Cao and Z.X. Cao, Appl. Phys. Lett. 83, 3474(2003).
A8. L. Pavesi et al., Nature 408, 440(2000).
A9. L. Dal Negro et al., J. Appl. Phys. 96, 5747(2004).
A10. S. Boninelli et al., MRS Symp. Proc. 817, L6.12.1(2004).
A11. M. Cazzanelli et al., J. Appl. Phys. 96, 3164(2004).
A12. P.G. Kik and A. Polman, Mater. Sci Eng. B81, 3(2001).
A13. D. Bertram et al., MRS Symp. Proc. 771, L4.49.1(2003).
A14. V. Kveder et al., Appl. Phys. Lett. 84, 2106(2004).
A15. P.K.H. Ho et al., Nature 404, 481(2000).
A16. G.L. Frey, K.J. Reynolds and R.H. Friend, Adv. Mater. 14, 265(2002).
A17. D. Moses, R. Schmeckel and A.J. Heeger, MRS Symp. Proc. 771, L5.9.1(2003).
A18. M.A. Laurence et al., Physica E16, 376(2003).
A19. M. Zelsmann et al., Appl. Phys. Lett. 83, 2542(2003); J.M Ziebarth et al., Adv. Funct.
     Mater. 14, 451(2004); Y.R. Do et al., J. Appl. Phys. 96, 7629(2004).
A20. S. Coe et al., Nature 420, 800(2002).
A21. D. Bertram et al., MRS Symp. Proc. 771, L4.49.1(2003).
A22. R. Mueller-Mach et al., IEEE J. Sel. Top. Quant. Elect. 8, 339(2002).
A23. D.Y. Kondakov, J. Appl. Phys. 97, 024503(2005).
A24. E. Hadjoudis and I.M. Mavridis, Chem. Soc. Rev. 33, 579(2004).
A25. J.-W. Son et al., MRS Symp. Proc. 784, C10.8.1(2004).
A26. L.R. Dalton, MRS Symp. Proc. 817, L7.2.1(2004); J. Phys.: Condens. Matter 15,
     R897(2003).
A27. S.S. Lee et al., IEEE Microwave Guided Wave Lett. 9, 357(1999).
A28. D.H. Chang et al., IEEE Photon Tech. Lett. 12, 537(2000).
A29. H. Fetterman, THz Electron Proc. 102, 1998.
A30. A. Yacoubian et al., IEEE J. Sel. Top. Quant. Elect. 6, 810(2000).
A31. L.R. Dalton in Adv. Polymer Sci. 158, 1(2001); D. An et al., Appl. Phys. Lett. 76,
     1972(2000).
A32. L. Sun et al., Opt. Eng. 40, 1217(2001).

Problems

1. How would you attempt to alter the band gap in a thin film of a given III–V
   semiconductor?
2. Given a particular thin film luminescent device, what treatment offers the
   possibility of enhancing the luminescent efficiency?
3. Suppose that for some reason the thickness of layers in a SiGe/Si multilayer
   assembly is increased beyond the critical thickness in order to achieve strain
   relief. Why would this multilayer not be able to function efficiently as a light-
   emitting device?
4. Design a thin film wave guide deposition process and justify your choice.

5. Why would you expect that the polarization, and, hence, the electro-optic coefficients, of a ferroelectric thin film may depend upon (a) film stress, (b) texture, if polycrystalline or orientation, if monocrystalline?

6. Why would you expect the polarization of thin films to be insensitive to defect structure?

7. How are high electro-optic coefficients achieved in organic thin films?

8. Describe some characteristics of the structure for a material to be capable of providing a holographic record.

9. What is the structural characteristic that defines an excellent electrochromic material?

10. Which surface would provide the higher solar absorptance – a surface of tungsten deposited by PVD at room temperature or a surface of tungsten deposited by PVD along with simultaneous high energy ion beam bombardment during deposition?

CHAPTER IV

Mechanical Properties

Mechanical properties of thin films and the interface between film and substrate will be considered in this chapter. Among these properties are flow strength, fracture strengths of film and of interface between film and substrate, friction and resistance to wear, elastic moduli, stress relaxation and piezoelectric properties. The mechanical properties of thin films affect their behavior in a variety of applications. In particular, the plastic flow and fracture, that occur during the operation of conductor lines upon electromigration induced stress development, certainly are affected by the mechanical properties of the film as well as the mechanical constraints provided by the substrate and cladding; adhesion of film to substrate is affected by the film's mechanical properties as well as the substrate/film interface strength; wear of thin film head and recording media thin film overlayers depend upon their mechanical properties, etc.

1. Flow and tensile strengths of thin films.

1.1. Flow strength.

Microstructure has marked effects on the strengths of bulk materials. It would be expected that thin films exhibit the same effects on their various strengths. For example, for ductile bulk materials, the yield strength increases as the dislocation density increases, as the grain size decreases, as the solute content in solution increases (for most types of solute), and as the spacing between non-coherent particles dispersed within the bulk decreases. The yield strength of ductile thin films exhibits similar qualitative behavior with these structural variations *only when the thin films are not constrained by a substrate or other more rigid layer*. For the usual case of a film on a substrate it is the elastic modulus difference between the substrate and film that exerts control over the yield strength of the film overriding all other microstructural effects.

Chu and Barnett[1] have recently reviewed the contributions made by various individuals to the problem of determining the yield strength of a film that is constrained by a substrate of higher elastic modulus. In the literature, Koehler[2] is normally listed as the first to calculate the stress on a dislocation in a film due to its nearest image in an adjacent layer. However, as Chu and Barnett noted, others[3] solved this problem prior to Koehler. Since then there have been several calculations of the shear stress required to move or generate dislocations in a film constrained by an adjacent rigid layer. We shall make use of the results of the first of these calculations by Sevillano.[4] According to Sevillano in the case where the elastic modulus of the substrate greatly exceeds that of the film then the increment in the lower-limit critical shear stress required to move a pre-existing dislocation half-loop confined in the thin film relative to that for the bulk material is given by:

$$\Delta\tau = (2\alpha\mu b\cos\theta/h)\ln[h/b\cos\theta] \qquad (4.1)$$

where α is $1/4\pi$ for screw dislocations and $1/4\pi(1 - \nu)$ for edge dislocations, ν is Poisson's ratio, μ is the shear modulus, b is Burger's vector, h is the film thickness and $h/\cos\theta$ is the width of the dislocation glide plane in the film between interfaces. Should dislocations need to be generated for slip to occur then h in equation (4.1) is replaced by h/2.

We can compare the yield stress corresponding to equation (4.1) with experiment using data for Al films. For this comparison we note by Schmid's law[5] that the yield stress $\sigma = \tau/m$, where m is the Taylor factor that depends upon the angles made by the slip plane and Burger's vector relative to the tensile (or compressive) stress axis. Normally, Al films have a (111) texture, so that θ in equation (4.1) equals $19°28'$ and m = 0.272 from application of Schmid's law to the geometry of the slip system relative to the texture. We have evaluated equation (4.1) for the case of Al with $\mu = 28.5\,\text{GPa}$, b = 2.86 Å and the resulting values for the yield

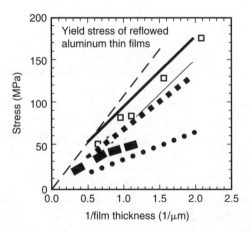

Figure 4.1. Data from Ref. [6] reproduced with permission. Heavy and light full lines correspond to the stress to generate or move edge dislocations in film, respectively. Heavy and light dotted lines correspond to the stress to generate or move screw dislocations in film, respectively. Light dashed line corresponds to the summary of experimental data due to Ref. [7]. The heavy dashed line corresponds to the yield stress of bulk aluminum on the assumption that the grain size equals the film thickness.

strength are plotted in Figure 4.1. The full lines correspond to edge dislocations and the dotted lines to screw dislocations. The heavy lines (full or dotted) correspond to the case that dislocations must be generated for slip to occur, whereas the lighter full or dotted lines correspond to the case where motion of existing half-loops initiate slip. The open square points represent experimental values taken from the work of Venkatraman.[6] The light dashed line in the figure corresponds to the relation given by Heinen et al.[7] as a summary of the experimental data for substrate bonded Al films from many different investigators. The heavy dashed line corresponds to the yield strength for bulk Al on the assumption that the grain size equals the film thickness and its dependence on reciprocal grain size is not linear. In the latter case, the yield stress varies inversely as the square root of the grain size, a relation known as the Hall–Petch relation with the yield stress in MPa given by $5.5 + 40.3/\sqrt{D}$, where D is in μm. Indeed, in thicker films ($>5\,\mu$m) the Hall–Petch relation applies. *It is apparent that the best fit between theory and experiment is for the case where the generation of edge dislocation segments limits the onset of crystal slip.*

The conclusion that the high yield strength of substrate bonded films is a consequence of the bonding of the film to a more rigid substrate is substantiated by experimental results for free-standing, unbonded films. Heinen et al.[7] and Lehoczky[8] have found that unbonded Al films have the same yield strength as bulk Al of the same grain size.

Multilayer films* may or may not be bonded to a substrate. If these multilayers consist of two ductile components then the yield strength of the multilayer depends upon the volume fraction and relative elastic modulus of each component according to the relation:

$$\sigma_m = R(f_1 + f_2 \cdot E_2/E_1)(\sigma + \sigma_\infty) \qquad (4.2)$$

where $R = (G_1 - G_2)/(G_1 + G_2)$, G_i are the shear moduli, f_i are the volume fractions of layers 1 and 2, E_i are the Young's moduli with $E_1 > E_2$, σ is the yield strength obtained by following the procedure described above, and σ_∞ is the yield strength of bulk material 2 of the same grain size as layer 2. This relation is derived on the basis that the applied elastic strain is the same in both layers and that yield of the laminate occurs when the product of the applied strain and Young's modulus of the lower modulus material exceeds ($\sigma + \sigma_\infty$), the yield strength of the constrained lower modulus material corresponding to its thickness. This result was found by Lehoczky[8] to hold for Al/Cu free-standing multilayers. The value of the slope of the yield strength versus reciprocal layer spacing obtained by Lehoczky[8] is nearly equal to that corresponding to the curve for the stress required to move screw dislocations in Figure 4.1, corrected for the factor $R(f_1 + f_2 \cdot E_2/E_1)$.

When the component of higher yield strength begins to yield then the dependence of multilayer yield strength upon wavelength with decreasing wavelength

---

* See Appendix for latest information concerning multilayers.

is somewhat problematical. Lehoczky found that the yield strength of Al–Cu multilayers remained constant from 70 nm down to 20 nm. For constant grain size it was observed that the nanohardness (proportional to yield strength) of Al/Ti multilayers constrained on a substrate increased with decreasing bilayer wavelength,[44] much as would be expected from the dependence of yield strength on this wavelength. The data collected by Hazzledine and Rao[42] for Cu/Ni multilayers implies a drop in yield strength with wavelength decreasing below 20 nm. Additional measurements of the yield strength of Cu/Ni multilayers[47] also revealed a maximum at a wavelength of about 20 nm. Similar results exhibiting maxima in nanohardness values, which are proportional to the yield strength, at wavelength values between 5 and 10 nm were reported for the single crystal multilayers: TiN/VN[52], TiN/V$_{0.6}$Nb$_{0.4}$N[53], TiN/NbN[54]. However, the multilayer VN/NbN[55] did not exhibit a maximum in the nanohardness-wavelength dependence over the same range of wavelengths.

An explanation of these results was provided by Chu and Barnett[51] based on the fact that the elastic moduli of VN and NbN are nearly equal whereas the other multilayer combinations involved pairs of materials having significantly different values of their shear moduli. They assumed, along with Krzanowski,[56] that the stress required to move a dislocation across the interface between the two component layers equals the value of the stress repelling the dislocation when it is in the center of the interface between the two layers. They evaluated this stress using prior work of Head[57] and Pacheco and Mura.[58] The result they obtained is that, at a layer thickness on the order of that observed (10–20 nm), the stress to move a dislocation across the interface between the layers becomes less than that necessary to move a dislocation in the layer of lower modulus and this stress decreases with decreasing layer thickness.

It has been found, for a substrate constrained Al film in which the grain size is much smaller than the film thickness, that the hardness follows a Hall–Petch dependence on grain size, i.e. hardness is linearly dependent upon the reciprocal square root of the grain size.[43] Thus, for the case where the grain size is less than the film thickness the Hall–Petch relation rather than equation (4.1) governs the yield strength. When the grain size is about equal to the film thickness Nix[59] concluded that the effect of grain size on the yield strength must be added to the effect of film thickness. (For normal deposition processes the average grain size is on the order of the film thickness. It is only when the film is subject to a sufficient fluence of high energy particles during deposition that the resulting average grain size may be much smaller than the film thickness. See the relevant discussion in I.*)

Plastic strain beyond the yield point of the ductile component needs to be accommodated by the other component. It is unlikely that a brittle component will

---

* Volume I in the series **Materials Science in Microelectronics**.

strain elastically to satisfy the plastic strain in the ductile component and it is more likely that the brittle component will fracture. The fracture aspect of this multilayer assembly will be considered in a section that follows.

1.2. Tensile strength.

There exists no quantitative theory to guide our examination of the effect of structure on tensile strength of thin films. Indeed, for a ductile film constrained to a substrate it is difficult to develop the strains that normally correspond to the tensile strengths of bulk materials in any type of test to which one can subject the substrate/film assembly. Hence, the only data on tensile strengths of thin films or multilayers of thin films come from tests on free-standing films or multilayers. Lehoczky[8] found that above a critical layer thickness of about 70 nm the tensile strength of Al/Cu multilayers varied according to the Hall–Petch relation with layer thickness, whereas below this critical thickness the tensile strength was constant down to a thickness of 20 nm. In another investigation, the tensile strength of Al/Ti multilayers exhibited a maximum at 20 nm for a variation between 10 and 110 nm.[45] Given the results described in the previous section on yield strength it seems likely that the dependence of tensile strength on the wavelength of multi-layer films represents a dependence of tensile strength on yield strength for these films.

2. Fracture in thin films bonded to substrates.

The thin film-substrate assembly may fracture via different modes. The most common mode is interface fracture or delamination which is considered in the next section. In this section, we will consider brittle fracture in brittle films, stress induced voiding and fracture during electromigration of conducting lines. Ductile fracture in ductile films usually does not occur because the stress in the latter type of film is limited by plastic strain of the film, providing that such plastic strain is homogeneous throughout the film. Should the plastic strain be inhomogeneous, as, for example by grain boundary sliding, then voids may be nucleated at the intersections of grain boundaries with the substrate, i.e. the phenomenon of stress voiding.[9] It is only during electromigration that a continuous build-up of stress, occasioned by a divergence in the flux of atoms, may occur. Such increase in the stress level may be limited in the case of tensile stress by an increase in the local vacancy concentration, by plastic strain, or by void formation and growth. In the case of compressive stress, a limit to its build up may be brought about by hillock extrusion. Also, there is the possibility that a stress gradient will develop of sufficient magnitude to bring the electromigration process to a halt. We will consider the failure modes during electromigration in greater detail later in this section.

2.1. Brittle fracture of substrate-bonded brittle films.

The fracture strength of brittle substrate-bonded films is a function of the defects present in the film and at the film/substrate interface. Films deposited at substrate temperatures below T_1 in zone 1 contain a grain boundary void network, which act as stress concentrations. Thus, the tensile stress to produce fracture of such films will be less than those of films deposited at substrate temperatures above T_1 or of films annealed at temperatures higher than T_1. However, the thermal stresses induced in the film may be higher in the films subjected to elevated temperatures and subsequently cooled to room temperature. Also, the enhanced mobility may allow the development of grain boundary grooves at the surface, and may allow grain boundary sliding to occur at the elevated temperatures. The consequences of both phenomena act as stress concentrations. Thus, prevention of brittle fracture of substrate-bonded films depends on the avoidance of stress concentrating defects and their formation, and of the build up of tensile stress in the film itself.

There are several strategies for enhancing the resistance to brittle fracture of non-ductile films. One possible solution is to arrange to have the film under a small compressive stress. However, too high a compressive stress can lead to buckling of the film away from the substrate via delamination at the film/substrate interface. Another possible technique for enhancing the resistance of brittle films to crack propagation is to use a ductile buffer layer between the brittle layer and the substrate. The function of the ductile buffer layer is to stop crack propagation from cracks emanating from the interface with the substrate before these cracks reach the brittle film. Indeed, experiments of Reimanis et al.[10] have shown that the use of increasingly thicker ductile Au layers between brittle Al_2O_3 layers increased the fracture initiation resistance and the subsequent crack growth resistance of the assembly. Further, the thinner is the brittle film the lesser will be the tendency to brittle fracture and delamination. Also, the use of a multilayer assembly, where possible, alternating thin brittle films with thicker ductile ones, may allow the multilayer film to perform the function of a brittle one of equal thickness, yielding a longer functioning life.[46,48] A third alternative is to control the structure of the brittle film, if possible. In this regard, the absence of a grain boundary void network, an equiaxed grain structure in place of a columnar grain structure, small grain size, a random texture of the polycrystalline film, and a distribution of second phases (precipitates) within the film can all yield greater resistance to crack propagation as well as limit the stress concentration factor due to blocked grain boundary sliding. The potential represented by control over the microstructure on brittle fracture of non-ductile films has been explored only briefly and, consequently, this field is ripe for study.

2.2. Thermal stress voiding.

Thermal stress voiding occurs during cool-down from an annealing temperature when the thermal expansion mismatch develops internal tensile stress in the film and the film relieves this stress by the nucleation and growth of voids. Structure determines the sites of void nucleation because nucleation will occur heterogeneously first at sites where the local free energy exceeds that of the intragranular material. For this reason, the favored void nucleation sites are grain boundary triple points and lines of intersection of grain boundaries with the substrate or capping layer. Both types of sites are activated, in part, because the usual operating mode of stress relief in the cool-down to room temperature in Al, the predominant interconnection metal, is grain boundary sliding. That is, at stresses below the yield strength, grain boundary sliding will occur in Al at a temperature of about 200°C.[11] A consequence of grain boundary sliding is that a stress concentration is developed at the intersection of the grain boundary with the constraining layer or an adjoining grain, which acts to increase the local free energy at this intersection.

Although void nucleation determines the physical position of the voids in the film the process limiting the rate of stress-voiding-type-failure of the interconnection is the growth rate of the voids and the latter is, in turn, limited by the rate of diffusion of vacancies to the voids. The driving force for such diffusion arises once a void is nucleated because the tensile stress normal to the void surface is zero. Thus, there is a gradient of the tensile stress from the void-free areas of the film to the void surface. This gradient in tensile stress is the driving force for vacancy migration to the void. Now, vacancies can diffuse faster along dislocations, grain boundaries, interfaces and surfaces then they can through the crystal lattice. Hence, structure is important in stress voiding also because it affects the rate of void growth. Dislocations, grain boundaries, interfaces and surfaces that intersect the voids are thus the active paths for diffusion of vacancies to these voids.

The structure represented by the extended defects just listed can vary enormously in films and interconnections. For example, consider narrow interconnections ($\sim$0.25 μm wide) compared to wider interconnections ($>$1 μm wide). The grain structure in the narrower interconnection consists of predominantly a bamboo-like structure in which one grain occupies the line width in any cross-section along the interconnection. In the wider line the grain structure is polycrystalline with several grains intercepting any cross-section of the line. One consequence of such a difference in grain structure is that the paths feeding vacancies to voids differ in the absence of grain boundary paths in the bamboo-like structure and the presence of these paths in the polycrystalline interconnection.

Hence, void growth should be slower in the narrower interconnection as compared to that in the wider line. Just this result was observed experimentally.[12a]

Even for the same line width, defect structure can vary between different interconnection lines. One aspect of this difference is the crystallographic texture. Most Al based interconnection lines have a non-ideal {111} texture. By "non-ideal" is meant that not all the grains have their (111) plane precisely parallel to the plane of the film. If the grains bounding a grain boundary had the ideal {111} texture then the grain boundary would belong to the tilt axis type of boundary in which the rate of self-diffusion normal to the tilt axis is much slower than along the tilt axis. Research has discovered that voids develop at boundaries between grains that do not have the ideal {111} orientation even though the majority texture is {111}, albeit under circumstances where the driving force for diffusion stems from the electron wind and not the gradient in tensile stress.[12b]

Another possible effect of structure on thermal stress voiding occurs in alloy films where the solid solubility of a component is a function of temperature. If the anneal temperature or temperature for the capping process is above the solvus line for the composition of the alloy then precipitation will occur at temperatures below the solvus line during the cool-down cycle. Thus, if the precipitate has a higher specific volume than the average component of the precipitate it is possible for precipitation to relieve a hydrostatic or biaxial tensile stress in the film. Whether this mode of tensile stress relief is a useful one depends upon the relative values of the thermal strain and precipitate induced counter strain, as well as volume fraction of precipitate. For example, in the alloy Al(2 wt.%Cu) $CuAl_2$ precipitates out of solution below the solvus temperature of about 400°C. Unfortunately, the crystal structure of this compound has a lower molar volume than the sum of two Al and one Cu atomic volume. Thus, precipitation acts to increase the tensile stress in the film of this alloy and exasperate the stress voiding. However, there is a positive contribution of the solute, copper, in that it reduces the stress voiding rate because by bonding vacancies it reduces the diffusion rate of vacancies along the extended defects. The net effect of the copper solute on stress voiding is probably a reduction of the stress voiding rate.

Another possible technique of reducing the tensile stress in a film constrained to the substrate and capping layers is to use what I will call stress-limiting, strain-compliant buffer layers. Let me consider the case of an Al film constrained to a silica substrate to illustrate what I mean by such a layer. Silica has a lower thermal expansion coefficient than Al ($8.3 \cdot 10^{-6}/°C$ versus $23.6 \cdot 10^{-6}/°C$). Hence, when Al cools down from its stress-free state at the anneal temperature (Al readily deforms plastically at this temperature) it will experience a tensile stress due to the constraint of the silica substrate. Now, if a buffer layer consisting of Ti–X wt.%Mo, where X lies between 7 and 9, is deposited between the silica substrate and the Al film it will have the bcc β structure and at room temperature it will be in a metastable state that can be transformed to the stable state, the hcp α structure, by the application of stress. The phase transformations, that are induced by the tractions applied to the buffer layer in cooling down from an annealing temperature, bring about an almost ideally plastic response of the buffer layer to these tractions.[13] Since the Ti based layer has about the same thermal expansion coefficient as silica ($8.4 \cdot 10^{-6}/°C$) the only traction applied to it will stem from the Al film. The buffer layer would then consist of gradients of one set of orientations of product phases that respond to the biaxial compression applied by the Al film. The volume percent of the α phase having this set of orientations would equal about 100% at the buffer layer/Al film interface and this percentage would decrease to zero near the interface between the buffer layer and substrate.

Finally, some mention should be made of the effect of dopants on the values of the grain boundary and surface energies. Decreasing the grain boundary energy and increasing the surface energy makes void nucleation more difficult and tends to make the voids more spherical in shape. (See E.S. Machlin, in **An Introduction to Aspects of Thermodynamics and Kinetics Relevant to Materials Science**, GiRo Press, 1991, Chapter IV.) This writer is not aware of any systematic study of this possible effect.

2.3. Electromigration modes of failure.[60]

Electromigration induces the building up of a tensile stress in one region of an interconnect line and a compressive stress in an adjacent region, along the direction of mass flow, to some maximum values limited either by void formation or hillock formation, respectively, or in the possible absence of these modes of failure, by a stress gradient that stops further electromigration. Usually, failure occurs before the stress gradient opposing the electromigration process can bring it to a stop. Tensile stresses develop in regions where the divergence of the matter flux is such as to build up a decrement in the matter density in the region and conversely, compressive stresses develop where the matter flux divergence results in an excess density in the region. *Thus, one effect of structure on failure in electromigration is the effect of the difference in number of grain boundary paths that carry electromigration induced matter flow into and out of a cross-section of an interconnect line, i.e. grain boundaries are preferred paths of matter flow because the self-diffusivity is higher along grain boundaries than within the grains.* This effect of structure on electromigration is illustrated by the example of a bamboo grain adjacent to a complex of several grains that occupy the cross-section of an interconnect line.

A capping layer on interconnect lines serves the function of hindering hillock formation. That is, before hillocks can proceed to grow the capping layer must fracture. *If we consider a capping layer to be an aspect of structure then its effect on the maximum compressive stress that can be attained in an interconnect line is another effect of structure on electromigration failure.*

Another effect of structure on electromigration failure is represented by the longest distance between a matter-deficient-divergence-source and the adjacent matter-excess-divergence-sink in the direction of matter flow, which we shall denote as the critical source-sink distance.* Since the tensile stress adjacent to a void is zero and since the strength of the capping layer determines the maximum compressive stress that can be developed at any cross-section of the interconnect line, *the value that the stress gradient that opposes matter flow can achieve, in a given capped interconnect line, is determined by the critical source-sink distance. The smaller is this distance the higher is the mass flow restricting stress gradient and the longer it takes to electromigration failure of the interconnect line.*

Conducting buffer layers are often used on either side of the interconnect line itself. These buffer layers are generally stronger than the interconnect line itself. The conducting buffer layers serve the functions of providing a conducting path shunting the current past open regions of the interconnect line where voids span an entire cross-section and of providing resistance to hillock formation. *Thus, conducting buffer layers represent another effect of structure on the useful life of an interconnect line.* (Shunt layers are not used for copper interconnect lines.)

The configuration at the ends of interconnections represents another effect of structure on electromigration failure. The end of a conducting line can connect to a via without any change in the interconnection width. Thus, a divergence in mass will be induced at the via/interconnection interface by the electron wind. In particular, at the cathode end of the interconnection, since the diffusivity of Al is much higher than that of W, a common via material, more Al will diffuse away from the original interface than W will diffuse through it. This type of via/interconnection geometry will lead to failure occurring by the opening of a void at the via/interconnection interface and will be independent of the microstructure and other parameters associated with the interconnection itself, as has been observed.[12a] This mode of electromigration failure can be ameliorated by the use of a pad of the same material as the interconnection, having a larger cross-section than the interconnection, and to which the via is connected.

When precipitates of an alloy interconnection occupy a complete cross-section of the latter then these precipitates act as sites of flux divergence, i.e. they block matter transport across them. Thus, the morphology and population density of such precipitates represent another effect of structure on electromigration. This effect can be good or bad. If the distance between blocking precipitates is larger than the critical source-sink distance defined above then the effect is bad because electromigration can continue to failure. However, if this distance is smaller than the critical source-sink distance then this effect is good because net mass transport is then nil.[61]

---

* This distance has been called the Blech length after Ilan Blech who first derived the equation relating the matter flux due to the stress gradient to the opposing mass flux driven by the electron wind.[60]

Indirect effects of structure related to the effects itemized above are line width (the smaller the line width the higher the fraction of bamboo-like grains, the smaller the average mass transport rate as a consequence of the reduction in grain boundary paths, and the smaller the average flux source-adjacent sink distance), and texture as affecting self diffusivity along grain boundaries. Similarly, the compressive stress that a capping layer can introduce into the interconnect line at the start of electromigration affects the time to build the compressive stress up to that at which the capping layer will fracture. Other variables are the fracture strength of the capping layer, its thickness, the electromigration resistance and mechanical strength of the buffer shunt layers, the difference in grain boundary diffusivity between different grain boundaries, etc.

Given the results described in Section 1.1 of this chapter and the analysis of the factors controlling failure of interconnect lines during electromigration presented above one potential means of enhancing the resistance to such failure is apparent. The stress level corresponding to the fracture that allows hillock production can be raised by taking advantage of the effect of film thickness on the yield strength and brittle fracture strength, i.e. use a multilayer array of conductor and higher modulus intermetallic compound shunt layers to provide the required low resistance of the interconnect with many thin conductor layers, each at the thickness corresponding to maximum yield strength. Not only should this architecture result in higher strength of the conductor layers, but the dispersion of the sites of mass divergences along the interconnect line should diminish their potential to bring about failure of the interconnect.

3. Interface fracture or delamination.

In this section we will consider briefly the fracture mechanics of the delamination process, the resistance of the interface to delamination, and the effect of interface structure on interface fracture.

3.1. Fracture mechanics of delamination.

Fracture occurs when the gradient of the energy driving crack propagation exceeds that resisting crack propagation. The latter energy is not an invariant for a given film/substrate material combination. It depends upon the loading that exists at the crack tip. For a crack that propagates along the interface between film and substrate the energy that resists crack propagation depends upon the interface structure, the ductility of the film and the inherent work of adhesion. For a crack that tends to propagate into the substrate it depends upon the fracture strength of the substrate and the existence of flaws in the substrate. For a crack that would tend to be driven into the film it depends upon the ductility of the film among other factors.

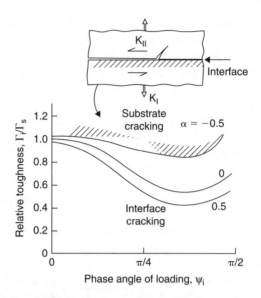

Figure 4.2. A map representing the crack path for given values of the relative fracture energies of interface to substrate and the phase angle of loading for different values of the parameter α, which increases as the shear modulus of material 1 increases relative to that of material 2. (Reprinted with permission from Evans et al., Acta Met. <u>37</u>, 3249(1989). Copyright 1989 Elsevier Science – NL, Amsterdam.)

Thus, an understanding of delamination requires a knowledge of the crack path activated by the stresses driving the fracture process.

Evans et al.[14] have provided a cogent summary of the fracture mechanics that govern the fracture path in bimaterial systems. Analysis of the factors that control the crack path shows that it depends upon the phase angle of loading Ψ. For the case of two brittle solids adjoining the interface which have the same shear modulus and Poisson's ratio values, the phase angle of loading is given by:

$$\Psi = \tan^{-1}(v/u) = \tan^{-1}(K_{II}/K_{I}) \qquad (4.3)$$

where u and v are the interface crack opening and crack shear, respectively, and K_{I} and K_{II} are the mode I and mode II stress intensity factors*, respectively. For systems where the shear moduli are unequal the relation for the phase angle depends upon the relative values of the moduli and Poisson's ratio.

For brittle bimaterial couples having unequal shear moduli, crack deflection out of the interface and into the lower modulus material will take place with the greatest tendency for such deflection when $\Psi = 70°$. Figure 4.2 provides a

---

* Mode I and II type stresses are defined in Figure 4.2 and correspond to tractions normal to and parallel with the interface, respectively.

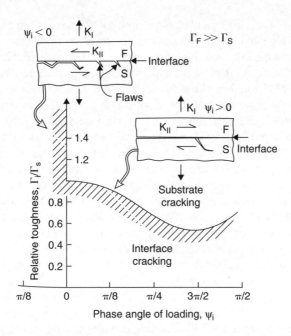

Figure 4.3. The region of fracture for a bimaterial system when one material, F, has a much higher fracture energy than the other, S. The result is plotted for the case where the moduli and Poisson's ratio of both materials are equal, respectively. When the phase angle of loading is less than zero segments of adjoining material can be attached to the stronger material. (Reprinted with permission from Evans et al., Acta Met. <u>37</u>, 3249(1989). Copyright 1989 Elsevier Science – NL, Amsterdam.)

guide to the path likely to be taken given a knowledge of the relative fracture energies of interface and lower modulus material and of Ψ. For the mode II shear arrows in the figure the lower modulus material is the top member. (α in the figure is positive when the bottom member is the higher modulus material.) Thus, even when the relative fracture strength of the interface is 0.6 that of the substrate the crack will extend into the latter when Ψ is 70° for the case that the two materials have the same moduli and Poisson's ratio, i.e. $\alpha = 0$.

If only one of the two materials (the substrate) is brittle and has much lower fracture energy than the other (the film) then crack propagation out of the interface into the brittle material will occur when the phase angle is positive. See Figure 4.3 for definition of positive phase angle for this case and for the case of equal moduli and Poisson's ratio. When the phase angle Ψ becomes negative, crack propagation out of the interface into the metal does not occur. Rather, for a low yield strength metal, once failure is initiated it proceeds along the interface when driven by sufficient force by ductile mechanisms involving hole nucleation at the interface. For a high yield strength material, the stress field of the interface crack can be high enough to cause cracks to propagate from flaws in the brittle material back to the interface resulting in pieces of brittle material being attached to the fracture surface, which in the main proceeds along the interface.[14] Thus, the phase angle of loading determines the crack path for a given bimaterial couple.

Delamination failure tends to initiate at edges and corners of the film/substrate interface where the interface intersects a surface and where large excess normal and shear tractions develop.[15] When these stresses are due to thermal expansion mismatch, the stresses decrease in magnitude with decrease in the film thickness. However, a mismatch due to modulus always generates interfacial tensile stresses at the edge whatever the sign of the mismatch and thus increases the tendency to fracture.[16] Data are provided in Ref. [16] to enable the reader to calculate the local stress field in the edge region given a knowledge of the differences in thermal expansion coefficient and temperature that determine the thermal strain component of the internal stress and the elastic moduli.

One mode of delamination is initiated by buckling of the film that starts at a region where the adhesion between the film and substrate is poor. Buckling occurs under the driving force of a compressive stress in the plane of the film, which may be there as a result of a difference in the thermal expansion coefficients between film and substrate. Once buckling is initiated, the buckle tends to propagate if the force driving the fracture exceeds the applicable fracture toughness resisting the fracture. Detailed studies and models have been made of the various modes of buckling.[17]

3.2. Resistance to delamination.

The initial resistance to delamination across an interface stems from the work of adhesion, which is the sum of the energies of the two surfaces produced by delamination along the interface less the energy of the interface between film and substrate. However, as mentioned in the previous subsection, the work of adhesion is only a small part of the total energy likely to be dissipated in interfacial fracture, at least for interfaces involving a ductile partner at the interface.

There are certain materials adjoining an interface for which fracture occurs along the interface with the work of adhesion the main resistance to the fracture. These are cases where the surface energies of the materials are very small, the interface energy relatively high and both materials are brittle. Different oxides having non-epitaxial interfaces tend to fall into this category. Also, metals bonded to brittle substrates, but for which the fracture occurs under stress corrosion conditions can fracture with the work of adhesion the main resistance to interface fracture.

When the film is ductile we have to distinguish between the case where the work of adhesion is small enough to allow for the development of only a limited number of "strong bridges" between substrate and film and the case where it is large enough to allow for unlimited bridging across the interface. The former case is represented by metals such as Au and Ag, which are relatively inert to oxygen and do not bond strongly to oxides, such as Al_2O_3. An empirical relation[18] that appears to govern the interface fracture energy when limited interface "bridging" is possible is:

$$\Gamma \approx W_{ad}(\sigma_o h/W_{ad} + 1)^{1/2} \qquad (4.4)$$

where W_{ad} is the work of adhesion, σ_o is the yield strength, and h is the thickness of the ductile film. Cu/SiO_2, Pt/Al_2O_3, and Nb/Al_2O_3 film/substrate combinations are among those that fit into this group of limited interface "bridging".

When a ductile material bonds strongly to the substrate in the absence of another phase between the two materials, then the work of adhesion becomes an insignificant parameter with respect to the resistance to delamination because of the onset of plastic flow. When void nucleation can occur along the interface and when the distance between voids is large compared to the film thickness the interface fracture energy is almost that of plastic dissipation at the yield strength and yield strain in the entire volume of the thin film. When the distance between voids becomes sufficiently small then the interface fracture energy is governed by this distance. Oh et al.[20b] have estimated the contribution to the interface fracture energy in this case to be proportional to the yield strength, the film thickness, the void diameter at the interface and inversely proportional to the distance between voids.

When a brittle phase forms between substrate and ductile film, as by reaction between the latter two materials, then it appears that interface fracture strength is related to the fracture strength of the brittle interphase. Amorphous interphases have interface fracture energies between 5 and $10\,J/m^2$, crystalline

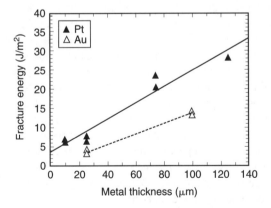

Figure 4.4. Trends in interface fracture energy with metal layer thickness for interfaces between Au or Pt and commercial grade Al_2O_3. (Reprinted with permission from Evans and Dalgleish, Acta Met. Mater. <u>40</u>, S295(1992). Copyright 1992 Elsevier.)

oxide interphase interface fracture energies range from 10 to 20 J/m^2, and those of crystalline intermetallic compounds between 20 and 40 J/m^2. Bridging to the ductile film also occurs leading to contributions from plastic flow to the interface fracture energy. Figure 4.4 indicates that for Au/Al_2O_3 and Pt/Al_2O_3 film/substrate systems the interface fracture energy depends linearly upon film thickness. Observations indicate that an amorphous layer forms at the interface in both systems. These results indicate that the interphase interface fracture energy is 6 J/m^2 or less, i.e. the intercept values in Figure 4.4.

The possibility that friction can be an additional process absorbing energy when the phase angle is negative has been treated by Stringfellow and Freund.[19] They found that the more elastically compliant the film and the higher the friction coefficient between film material and substrate material during sliding friction the greater will be the energy absorption hindering the delamination process.

Thus the main factors contributing to the resistance to delamination are the work to form new surfaces (brittle material couples), the energy dissipated in plastic flow (the main factor when at least one of the material couples is ductile), and the energy dissipated in frictional sliding when the phase angle of loading is negative.

3.3. Effect of interface structure on delamination.

Interface structure can have a significant effect on the effective work of adhesion. For the case of metal/oxide bonds, it is well known that the oxide surface terminates on the oxygen atoms. These atoms can be removed by subjecting the surface to an argon ion beam of a few hundred eV and maintaining the ion beam during deposition of the metal film. In this case, many of the metal atoms depositing onto the oxide surface will bond to the metal component of the oxide. If this bond is stronger than the corresponding bond to a surface oxygen atom of the oxide then the work of adhesion of the resulting interface will be higher than of that formed between the metal and the oxygen terminated oxide surface. Such results have been found for Cu on many oxides.[20a] Further, subjecting such an interface to either an annealing cycle sufficient to allow mass transport of the metal atoms at the interface or to a high energy irradiation of sufficient fluence to achieve ion mixing at the interface will result in an increase in the work of adhesion due to an increase in the metal–metal bond density across the interface. Obviously, this effect of bond type and density at the interface depends upon the relative bond strengths, i.e. $metal_{film}$–$metal_{oxide}$ versus $metal_{film}$–oxygen or $metal_{oxide}$–oxygen. Thus, for a Ti film it is not obvious that removal of the surface oxygen atoms of the oxide substrate would result in an increase in the work of adhesion.

Interface structure can have a significant effect also on interface fracture energy and on the delamination process. In the regime where void nucleation occurs during interface fracture and the distance between void nuclei directly affects the interface fracture strength it has been found that voids are nucleated at three-grain junctions of the substrate microstructure at the interface. Thus, according to the suggestion of Oh et al.[20b] refining the grain size should increase the interface fracture energy. Oh et al.[20b] have shown that deliberate patterning of "void" sites along an interface between Cu and SiO_2, with Cr used as a bond layer away from the void sites, enhances the interface fracture energy significantly.

Microstructure affects the interface fracture energy in many types of bimaterial couples involving a ductile film because this energy depends upon the yield strength of the constrained film and the latter depends upon crystallographic texture, as discussed in Section 1 of this chapter. Further, the "bridges" between voids are unconstrained and their mechanical properties, such as yield strength and yield strain, should depend sensitively on microstructure, as do these properties in bulk materials.

We have already noted that the interface fracture energies of interphase materials along interfaces depend upon whether the phases are amorphous or crystalline and whether the latter are oxides or intermetallic compounds. Thus, depending upon the circumstances there may or may not exist the possibility to control the type of interphase that develops between film and substrate so as to maximize the interface fracture energy. For example, when the interface fracture energy is low indicating poor adhesion between film and substrate a possible solution is to ion mix the interface to produce an interphase between the two materials that is likely to be an amorphous phase.

4. Wear and frictional properties of films.

The applications of thin films as friction and wear resistant materials are so varied as to make a generic discussion of the effect of thin film structure on the tribological properties difficult. However, there are certain generalizations that hold for all applications. According to Bowden and Tabor[21] the friction force can be divided into two parts: a plowing term for the force needed to move the asperities of the harder material through the softer material, and a shearing term for the force needed to shear the contacting junctions. From this concept we may deduce some of the factors that affect wear. Ploughing may involve not only the pushing apart by plastic flow of material in the path of sliding harder asperities, but may also involve the cutting out of material, much as a machine tool does in a typical machining operation. Thus, one type of wear is that produced by asperity machining. Should the junctions at the contacts between surfaces involve sufficient adhesion then when sliding occurs the concomitant shearing process is through the weaker material with a fragment of the weaker material attached to the stronger one at its point of contact. Hence, another type of wear involves welding contact and shear induced fracture of a portion of the weaker material, which we will call contact fracture.

Except for very light normal forces between the sliding surfaces, when elastic deformation controls the contact area, the latter is determined by the yield strength of the softer material and the normal force via the relation, $A = P/Y$, where P is the normal force and Y is the yield strength. Thus, the harder the materials in sliding contact the smaller will be the contact area and the lower should be the wear from both asperity machining and contact fracture. This conclusion is consistent with experience and accounts for the fact that most anti-wear thin film coatings consist of hard materials. Of course, other factors affect the wear resistance, such as the morphology of the contact fracture. For thin films there are various modes of contact fracture including spalling fracture, brittle fracture, and ductile shear fracture. Each of these modes produces a build-up of material at a contact point that ultimately separates from it by fracture producing wear particles. The latter can act as abrasive particles enhancing the wear process. Resistance to this abrasive wear increases with hardness of the sliding surfaces. However, increasing the hardness of a given material does not necessarily lead to an increase in its wear resistance if the process of increasing the hardness involves a change in the morphology or ease of contact fracture of the material.

Of course not all contacts need be welded contacts. Whether solid state welding occurs or not depends to a great extent upon the work of adhesion. As discussed in the section on delamination there are certain material couples for which the work of adhesion governs where the fracture occurs. A surface that is covered by a monolayer of sulfur or oxygen atoms is less likely to weld to another surface that consists of atoms that bond weakly to sulfur or oxygen, e.g. Au. Thus, the fracture in this case would be along the interface and not likely to produce any wear fragments.

From the simplified discussion of the wear process given above we may induce how structure may affect it. However, the anti-wear film is part of a materials system determined by the application. For example, consider a magnetic recording disk and head assembly. In this case, there are very thin anti-wear films between disk and head and normal contact between the two can lead to plastic deformation in the substrates just beneath the thin anti-wear films. Thus, not only do the properties of the anti-wear films affect the wear resistance, but the plastic flow properties of the immediate substrates also may affect the wear resistance by affecting the stresses induced in the anti-wear film and thereby its spalling resistance.[22]

One coating that should lead to a low friction and wear, and where the structure of the coating is designed to be consistent with the fundamentals governing these properties as discussed above, is a composite coating consisting of small hard particles dispersed and bonded to an ideally plastic matrix (one that does not work harden), which is itself strongly bonded to a harder substrate. The function of the hard particles is to limit the area of contact between the sliding surfaces and prevent penetration of the mating surface down to the substrate of the coating. Thus, it must be present in a significant volume fraction. The function of the ideally plastic matrix is to make the plastic flow process in the matrix a lower energy process than that involved in asperity machining and to prevent the formation of work hardened wear particles that can act as abrasive wear particles. Instead, wear particles replenish the matrix of the coating and act to coat hard surfaces and maintain the function of the ideally plastic material, i.e. provide a low shear resistant material. One example of such a coating is a composite coating consisting of Mo particles dispersed in a Ag matrix.[23] However, if the substrate to be protected from wear is itself a very hard material, then use of a thin ideally plastic material as a coating without interspersed hard particles can serve to reduce the friction and wear by the mechanisms detailed above.[24] In effect, the ideally plastic material acts as a solid lubricant separating the substrate from the sliding partner.

Assume that the hard anti-wear film is strongly bonded to its substrate so that it will not spall. One effect of structure, where such films are crystalline, brittle, and fracture intergranularly, is on the size of the contact fracture fragment. Obviously, under such circumstances, the smaller is the grain size, the smaller will be the contact fracture segment, and the smaller the wear. Further, under these circumstances, but where fracture occurs intragranularly, one would expect the resistance to contact fracture to increase the smaller is the grain size and the more random is the crystallographic texture of the film. As an example of the beneficial effect of a small grain size, a uniform equiaxed morphology and a random crystallographic texture, diamond films of very fine grain size (average grain size of 15 nm) having these characteristics were found to have two orders of magnitude smaller wear rates than diamond films having a microstructure consisting of a mixture of small equiaxed grains (8 nm average grain size) and much larger dendritic grains (longer than 110 nm).[49] Another example, which this writer believes exhibits this effect of size of contact fracture unit on wear rate, is that where a multilayer (TiN/Ti) wears less than single layers of either component of the same total thickness.[50] The films and layers in the multilayer film were amorphous in structure. In the multilayer film the size of the contact fracture fragment is controlled by interface fracture (i.e. one dimension of the fragment is the thickness of a single layer). In the single layer film one dimension of the contact fracture segment is the total layer thickness. Hence, the greater the number of layers in the multilayer, for a given total thickness, the smaller is the small dimension of the contact fracture segment. This analysis suggests that use of crystalline multilayer assemblies of hard and ductile layers should yield lower wear rates than a single film of the hard layer of the same total thickness which may even be less than that obtained for a multilayer of amorphous allotropes of these same materials.

It is apparent that the effect of structure on friction and wear will depend upon the particular materials couple because of the multitude of failure modes in contact fracture, i.e. spalling, brittle or ductile fracture, inter- or intragranular fracture path. Thus, one objective in any investigation of the friction and wear of a given sliding materials system is to determine the mode of contact fracture.

5. Elastic moduli.

The anomalous values of elastic moduli reported in Cu/Ni[25], Cu/Pd[26], Ag/Pd[27], and Au/Ni[26] thin film multilayers have since been shown by more sophisticated measurement techniques to be non-existent or much smaller in magnitude.[28] In bulk materials elastic moduli are insensitive to microstructure. It is now apparent that elastic moduli in thin films are also insensitive to microstructure except in films thin enough for the interface energy to represent a significant fraction of the total free energy difference between competing structures of at least one component of the multilayer. Among the structures that can compete for existence in multilayers are: different crystalline structures, one of which is metastable and the other stable in bulk; amorphous versus crystalline structures; coherently strained at a coherent interface versus unstrained structures at an incoherent interface.

It appears that a softening in the C_{44} shear elastic modulus occurs in certain *immiscible* pure metal bcc(110)/fcc (111) multilayer systems.[29] These are multilayers where the difference in atomic radii between the component metals are large. In this case it has been found that at least one of the component layers becomes amorphous.[33] These observations have been rationalized on the basis that the total free energy difference between two configurations of the system is nil.[30] These two configurations are: (1) the case where the phases on either side of the interface are stable and the interface energy is high due to an appreciable misfit in area per atom across the interface and (2) that where one of the phases assumes a metastable structure such that the interface energy diminishes (The considerations are very similar to those for the pseudomorphic stabilization of a metastable phase – see I.).

In this case, it appears that the metastable structure has a lower shear modulus than the stable structure and it rather than the interface provides the measured decrease in shear modulus. This interpretation is supported by the fact that the decrease in modulus is not proportional to the number of interfaces per unit length normal to them, but is proportional to the volume of the phase that becomes metastable.[31]

Incidentally, it is well known that in a hard sphere model the bcc structure is unstable with respect to a $(1\bar{1}0)[111]$ shear.[32] Thus, one might be tempted to suggest that the observed shear modulus softening for a system involving a bcc(110) plane parallel to the interface is related to this characteristic of a bcc structure. However, to use this explanation one would have to provide a reason why the bulk bcc structure does not suffer this shear softening while the thin film does. The explanation given by Grimsditch et al.[30] that the energy of the interface drives a transformation in a sufficiently thin film to a metastable structure having a lower shear modulus is much more reasonable. The lower shear modulus structure appears to be a disordered one in some cases, as revealed by X-ray diffraction.[33]

As indicated above, pseudomorphic stabilization of a metastable crystalline phase at the expense of an interface energy difference can occur. In this case, the measured elastic modulus at small multilayer wavelengths senses the metastable phase, while at large wavelengths it senses the stable phase.[34]

Finally, at small wavelengths it is possible to maintain coherent interfaces at the expense of elastic strain in the adjoining layers, while beyond a critical wavelength interface coherency is lost and the elastic strain reduced by misfit dislocation generation. In this case, elastic constant measurements sense two different configurations (elastically strained versus unstrained) in the layers and may yield different values for the two cases.[35]

When the two different layers in a multilayer assembly are miscible or form compounds then the changes in elastic moduli that occur are most likely a manifestation of the difference in elastic modulus between the interface phase, the interphase, and the mean of those replaced by this interphase.

Thus, when the crystallographic structure, state of strain, or composition of a material is changed in the process of forming a multilayer we can say that the elastic modulus of the multilayer is structure sensitive.

6. Stress relaxation.

The constraint of a substrate affects not only the stresses developed in deposited thin films, but it also affects the modes and consequences of stress relaxation in thin films as well. Stress relaxation involving sliding, whether along slip planes or along grain boundaries, will develop stress concentrations at the intersection of the slip plane or grain boundary with the substrate. Also, because of the substrate constraint, it is unlikely that stress can be uniformly relaxed throughout the film. Rather, it is more likely that a gradient will develop in the magnitude of the stress that is relaxed, with the largest relaxation occurring at the film surface and least at the substrate surface. Stress relaxation in thin films differs to another extent from bulk materials, because the grain size is much smaller in thin films than in bulk materials, and because stress in much of the thin film can be relaxed by mass transport out of the free surface.

The stress relaxation mechanisms that operate in thin films are the same as those that operate in bulk materials. Further, the effects of structure on these mechanisms are the same in thin films as in bulk phases, except where certain structures present in or processes associated with thin films are absent in bulk phases. For example, multilayers consisting of miscible pure component films can relax stress by interdiffusion under certain circumstances; implantation of inert gas atoms into thin films can relax tensile stresses in the films; ion bombardment of sufficient fluence can produce enough thermal spikes to bring about stress relaxation throughout most of a thin film.

Frost and Ashby[36a] have provided deformation mechanism maps for a variety of materials. These are maps of normalized stress (stress/modulus) versus homologous temperature (temperature/melting temperature). We find that Coble creep (atom transport from interfaces normal to compressive stress to interfaces normal to tensile stress *along grain boundary paths*) is the preferred mechanism at low normalized stress and homologous temperature. For homologous temperatures above about 0.5 and low normalized stress the preferred mechanism is Nabarro–Herring creep, which involves the same driving force for creep as the Coble mechanism, but the path for atom transport is within the lattice. Again above a homologous temperature of about 0.5, but at higher stress, creep occurs via dislocation climb. At still higher normalized stress and at all temperatures dislocation glide becomes the preferred creep mechanism. Since stress relaxation is creep occurring at a constant strain, but decreasing stress, we may expect that these mechanisms also operate for stress relaxation.

There exists the possibility that another mechanism of stress relaxation may operate in thin films, but not in bulk materials. This mechanism is grain boundary sliding with or without concomitant deformation of the adjoining grains. Such sliding would give rise to rotation of columns of grains and the development of stress concentrations at the substrate. The latter would tend to bring this mode of relaxation to a halt unless voids are nucleated or the grains deform.

The formation and growth of hillocks is a manifestation of the existence of compressive stress in the film. The propensity for hillocks to appear on thin films is probably a consequence of the fine grain size prevalent in thin films. Thus, it is likely that Coble creep is the relaxation mechanism responsible for most of the hillocks observed on thin films. The creep rate under the Coble mode is:

$$\dot{\varepsilon} = 10\sigma\Omega\pi\delta D/L^3 kT \qquad (4.5)$$

where σ is the stress, Ω is the atomic volume, δ is the grain boundary thickness, D is the grain boundary self-diffusivity, L is the grain diameter, k is Boltzmann's constant, and T is the absolute temperature. Thus, the structure dependence of stress relaxation in the Coble mode is given by the grain diameter L and by the grain boundary self-diffusivity, which may well be a function of the crystallographic texture of the film. In the Nabarro–Herring mode L^3 in the above equation is replaced by L^2 and the applicable diffusivity is that for lattice self-diffusion. When the grain size is large surface diffusion may contribute significantly to the overall transport. Further, when the grains are columnar in shape Thouless[36b] showed that L^3 in equation (4.5) is replaced by L^2h, where h is the film thickness, if grain boundary diffusion controls and by Lh^2 if surface diffusion controls.

Hillock formation is undesirable for many reasons. As one example it leads to a rough surface that is unacceptable as a substrate for further deposition. In electromigration circumstances hillock formation can lead to the production of electrical "shorts" and failure. Thus, elimination of hillock formation is desirable. There are several paths that may lead to this objective. Since compressive stress is the driving force for hillock formation then its reduction or removal should be helpful. As noted above, grain size increase can lower the rate of hillock formation. Further, decrease in the self-diffusivity along the grain boundaries that supply the hillocks with matter should also act to lower the rate of hillock formation. Any change in crystallographic texture, such as that of replacing large angle boundaries with small angle ones, or alloying that yields grain boundaries having lower self-diffusivities should be helpful as well, as was found to occur when small amounts of Si and Cu were added to Al.[37] However, the knowledge of how to achieve a desired texture in a commercial setting or of which alloying elements will act to reduce self-diffusivities along grain boundaries is, very likely, not yet available. (See, however, Ref. [38] for the effect of alloying elements on self-diffusivities along grain boundaries.)

When dislocations are involved in stress relaxation, microstructure can affect the stress relaxation rate significantly, just as microstructure affects the creep rate in bulk materials in the dislocation control regimes of the deformation-mechanism maps. For such effects of microstructure see any text on creep, such as Ref. [39].

7. Piezoelectric and surface acoustic wave properties.

Non-ferroelectric polycrystalline films exhibit piezoelectric properties only when they are deposited to produce the required crystallographic texture. Among such materials are ZnO, Ta_2O_5, and AlN. For these materials, the piezo-electric properties improve with greater perfection (lower deviation) of the pre-ferred orientation, and with larger grain size. Figure 4.5 reveals that the surface acoustic wave (SAW) propagation loss in a ZnO film decreases with decrease in the etch pit density. The latter quantity is denoted as the defect density and is likely to be the dislocation density. Also shown in Figure 4.5 is the loss for propagation of guided acoustic waves, which appears to be much more sensitive to the defect density. An interesting result is shown in Figure 4.6. The piezoelectric coupling factor that measures the efficiency of conversion of electrical to mechanical energy does not become significant until the distance between defects exceeds the acoustic wavelength.

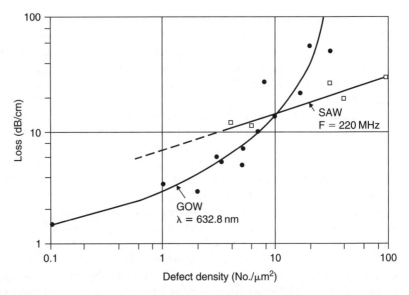

Figure 4.5. Dependence of propagation loss on etch pit density (dislocation density). (From F.S. Hickernell, MRS Symp. Proc. <u>47</u>, 63(1985) with permission, see Ref. [40].)

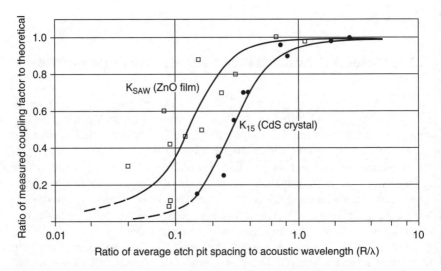

Figure 4.6. Piezoelectric coupling factor dependence on wavelength normalized defect spacing. (From F.S. Hickernell, MRS Symp. Proc. <u>47</u>, 63(1985) with permission.)

For ferroelectric based materials, the piezoelectric capability is related to domain switching which may be hindered by the substrate constraint. In bulk materials over half of the piezoelectric response is due to domain wall motion and this motion is increasingly hindered as the grain size decreases. Since the grain size of thin films is normally small it may be expected that the piezoelectric response of thin films would be less than for bulk samples of lead zirconium titanate (PZT). However, the piezoelectric properties of PZT polycrystalline films at the present writing are sufficient to have stimulated a host of microelectro-mechanical system (MEMS) developments. Further, the successful switching of domains of PZT thin films in DRAM application also suggests that although switching of domains may be more difficult than in bulk materials it is not completely quenched in thin films, i.e. there are other mechanisms of domain reorientation than domain wall motion, such as via the nucleation of new domains and rotation. Indeed, d_{33} coefficients ranging from 150 to 500 pC/V have been reported[62,63] for thin films of PZT, which suggests that microstructure may affect the piezoelectric properties of thin films. At the time of this writing nothing has appeared in the literature concerning the effect of microstructure on the piezoelectric properties of thin films other than that improved piezoelectric characteristics appear with increase in film thickness.

8. Mechanical fatigue properties.

There are few data regarding the mechanical fatigue behavior of thin films bonded to substrates. One such investigation revealed that the crack growth rate was larger than would normally occur in bulk for the metal component.[41] It was suggested that a possible reason for this behavior is that larger amplitude shear stress reversals occur near the crack tip when it is located near the metal/substrate interface. For a microlaminate of Cr_2Nb/Nb the fatigue crack growth rate was only slightly higher than for pure Nb.[46] Apparently, fatigue crack propagation in the multilayer ceramic/metal structure is mainly controlled by the metal layer. Also, the structure dependence of the fatigue resistance for the metal film component is likely to be the same as it is in the bulk metal. However, see Section A1.2 of Appendix.

Recapitulation

The yield strength of thin films thicker than 20 nm and constrained by a substrate of higher elastic modulus corresponds to the stress required to generate edge dislocations in the thin film. For thin films constrained by a substrate and thinner than 20 nm the yield strength corresponds to the stress required to move dislocations across the interface into the higher modulus substrate. The same considerations apply to thin film multilayers. However, for grain sizes equal to or smaller than the film thickness the effect of grain size on the yield strength must be added to the values calculated according to the previous considerations.

Tensile strength scales with the yield strength of ductile thin films. The fracture strength of brittle thin films is a function of the stress concentrating defects present in the films. Multilayering such that brittle layers are separated by ductile ones enhances the crack initiation and growth resistance of the brittle components. Tensile stress induced void formation is enhanced at stress concentrations due to grain boundary sliding in constrained thin films. Structure affects void formation by either reducing the level of tensile stress or by enhancing the resistance to grain boundary sliding.

Failures along interconnect lines during electromigration are induced by stresses developed at matter flux divergences. Little is known about how to control the rate of build-up in the magnitudes of these divergences. The best hope of enhancing the resistance to electromigration induced failure is to increase the strength levels corresponding to the onset of hillock growth and, thereby, the stress gradients opposing the electromigration induced mass flow. A potential means of accomplishing this objective is the use of a multilayered interconnect line involving thinner conductor layers interspersed with shunt type strong intermetallic compound layers.

Stresses in thin films tend to bring about delamination. The main factors contributing to the resistance to delamination are the work to form new surfaces (brittle material couples), the energy dissipated in plastic flow (the main factor when at least one of the material couples is ductile), the energy dissipated in nucleating and growing ductile fracture voids at the interface, and the energy dissipated in frictional sliding when the phase angle of loading is negative. The interface structure is capable of being modified and by this means it is possible to control the resistance to delamination of certain film/substrate couples.

Asperity machining and contact fracture are the two main modes contributing to wear of sliding surfaces. Also, wear particles, once produced and acting as abrasive particles, can enhance the wear process. This knowledge together with a knowledge of the likely modes of contact fracture enables the design of an optimum anti-wear film/substrate system.

Differences in interface energy in thin film multilayers may drive the formation of metastable phases that have lower elastic modulus than the stable phases. Also, intermixing at interfaces can lead to an interface interphase that has an elastic modulus different from the bounding phases.

Stress relaxation in thin films by mass transport out of the free surface can occur more readily than in bulk materials. Hence, hillock formation is a problem in many circumstances involving the use of thin films.

Dislocations can deteriorate the coupling efficiency for SAWs when the spacing between dislocations is less than the acoustic wavelength.

References

1. X. Chu and S.A. Barnett, MRS Symp. Proc. 382, 291(1995).
2. J.S. Koehler, Phys. Rev. B2, 547(1970).
3. A.K. Head, Phil. Mag. 44, 92(1953); E.S. Pacheco and T. Mura, J. Mech. Phys. Solid.. 17, 163(1969).
4. J.G. Sevillano, in **Strength of Metals and Alloys**. Proc. ICSMA 5, eds. P. Haasen, V. Gerold and G. Kowtorzs, Pergamon Press, Oxford, 1980, p. 819.
5. See, for example, G.E. Dieter, in **Mechanical Metallurgy**, McGraw-Hill Book Co., New York, 1986.
6. See data due to Venkatraman in S.P. Baker and W.D. Nix, SPIE, 1323, **Optical Thin Films III: New Developments**, 1990, p. 263.
7. D. Heinen, H.G. Bohn and W. Schilling, J. Appl. Phys. 77, 3742(1995).
8. S.L. Lehoczky, J. Appl. Phys. 49, 5479(1978).
9. M.A. Korhonen, W.R. LaFontaine, P. Børgesen and C.-Y. Li, J. Appl. Phys. 70, 6774(1991).
10. I.E. Reimanis, B.J. Dalgleish and A.G. Evans, Acta Metall. Mater. 39, 3133(1991).
11. P. Børgesen, M.A. Korhonen, C. Basa, W.R. LaFontaine, B. Land and C.-Y. Li, MRS Symp. Proc. 225, (1991).
12. (a) C.-K. Hu, K.P. Rodbell, T.D. Sullivan, K.Y. Lee and D.P. Bouldin, IBM J. Res. Develop. 39, 465(1995); (b) J.L. Hurd, K.P. Rodbell, D.B. Knorr and N.L. Koligman, MRS Symp. Proc. 343, 653(1994).
13. P. Børgesen, M.A. Korhonen, T.D. Sullivan, D.D. Brown and C.-Y. Li, MRS Symp. Proc. 239, 683(1992).
14. A.G. Evans, B.J. Dalgleish, M. He and J.W. Hutchinson, Acta Met. 37, 3249(1989).
15. A.G. Evans, M. Ruhle and M. Turwitt, J. de Physique 46, C4–C613(1985).
16. T. Suga and G. Ellsner, in **Proceedings of the MRS International Meeting on Advanced Materials Vol. 8 – Metal–Ceramic Joints**, eds. M. Doyama et al., 1989, p. 99.
17. M. Ortiz and G. Gioia, J. Mech. Phys. Solid. 42, 531(1994); J.W. Hutchinson, M.D. Thouless and E.G. Liniger, Acta Metall. Mater. 40, 295(1992).
18. A.G. Evans and B.J. Dalgleish, Acta Met. Mater. 40, S295(1992).
19. R.G. Stringfellow and L.B. Freund, MRS Symp. Proc. 239, 579(1992).
20. (a) J.E.E. Baglin, MRS Symp. Proc. 47, 3(1985); (b) T.S. Oh, J. Rödel, R.M. Cannon and R.O. Ritchie, Acta. Metall. 36, 2083(1988).
21. F.P. Bowden and D. Tabor, in **The Friction and Lubrication of Solids**, Oxford University Press, New York, 1950.
22. H.-F. Wang, J.C. Nelson, C.-L. Lin, J.W. Hoehn and W.B. Gerberich, MRS Symp. Proc. 356, 761(1995).
23. Y.-T. Cheng, B. Qiu, S. Tung, J.P. Blanchard and G. Drew, ibid., p. 875.
24. J.H. Hsieh, O.O. Ajayi, A. Erdemir and F.A. Nichols, MRS Symp. Proc. 239, 629(1992).
25. T. Tsakalakos and J.E. Hilliard, J. Appl. Phys. 54, 734(1982); D. Baral, J.B. Ketterson and J.E. Hilliard, J. Appl. Phys. 57, 1076(1985).
26. W.M.C. Yang, T. Tsakalakos and J.E. Hilliard, J. Appl. Phys. 48, 876(1977).

27. G. Henein and J.E. Hilliard, J. Appl. Phys. 54, 728(1983).
28. A. Moreau, J.B. Ketterson and B. Davis, J. Appl. Phys. 68, 1622(1990).
29. I.K. Schuller, A. Fartash and M. Grimsditch, MRS Bull. XV, 33(1990); A. Fartash, I.K. Schuller and M. Grimsditch, Rev. Sci. Instrum. 62, 494(1991).
30. M. Grimsditch, E.E. Fullerton and I.K. Schuller, MRS Symp. Proc. 308, 685(1993).
31. M.R. Khan, C.S.L. Chun, G.P. Felcher, M. Grimsditch, A. Kueny, C.M. Falco and I.K. Schuller, Phys. Rev. B27, 7186(1983).
32. C. Zener, in **Elasticity and Anelasticity of Metals**, University of Chicago Press, Chicago, 1965.
33. I.K. Schuller and M. Grimsditch, J. Vac. Sci. Tech. B4, 1444(1986); R. Danner, R.P. Huebner, C.S.L. Chun, M. Grimsditch and I.K. Schuller, Phys. Rev. B33, 3696(1986).
34. S.F. Cheng, A.N. Mansour, J.P. Teter, K.B. Hathaway and L.T. Kabacoff, Phys. Rev. B47, 206(1993).
35. J.R. Dutcher, S. Lee, J. Kim, J.A. Bell, G.I. Stegman and C.M. Falco, Mat. Sci. Eng. B6, 199(1990).
36. (a) H.J. Frost and M.F. Ashby, in **Deformation-Mechanism Maps**, Pergamon Press, Oxford, 1982; (b) M.D. Thouless, Acta Metall. Mater. 41, 1057(1993).
37. C.M. Su, H.G. Bohn, K.-H. Robrock and W. Schilling, J. Appl. Phys. 70, 2086(1991).
38. I. Kaur and W. Gust, in **Handbook of Grain and Interface Boundary Diffusion Data**, Ziegler Press, Stuttgart, 1989.
39. F.R.N. Nabarro and H.L. de Villiers, in **The Physics of Creep: Creep and Creep-Resistant Alloys**, Taylor & Francis, London, 1995.
40. F.S. Hickernell, MRS Symp. Proc. 47, 63(1985).
41. A.G. Evans and J.W. Hutchinson, Acta Metall. Mater. 43, 2507(1995).
42. P.M. Hazzledine and S.I. Rao, MRS Symp. Proc. 434, 135(1996).
43. J.B. Savader, M.R. Scanlon, R.C. Cammarata, D.T. Smith and C. Hayzelden, MRS Symp. Proc. 403, 157(1996).
44. R. Ahuja and H.L. Frasier, J. Met. 46(10), 35(1994).
45. D. van Heerden, D. Josell and D. Shechtman, MRS Symp. Proc. 403, 139(1996).
46. J. Heathcote, G.R. Odette, G.E. Lucas and R.G. Rowe, MRS Symp. Proc. 434, 101(1996).
47. T. Nakano, A. Yokoyama and Y. Umakoshi, Scripta Metall. Mater. 27, 1253(1992).
48. D.R. Bloyer, K.T. Venkateswara Rao and R.O. Ritchie, MRS Symp. Proc. 434, 243(1996).
49. R. Csencsits, D.M. Gruen, A.R. Krauss and C. Zuiker, MRS Symp. Proc. 403, 291(1996).
50. Ph. Houdy, P. Psyllaki, S. Labdi, K. Suenaga and M. Jeandin, MRS Symp. Proc. 434, 57(1996).
51. X. Chu and S.A. Barnett, J. Appl. Phys. 77, 4403(1995).
52. U. Helmersson, S. Todorova, S.A. Barnett, J.-E. Sundgren, L.C. Markert and J.E. Greene, J. Appl. Phys. 62, 481(1987).
53. P.B. Mirkarimi, L. Hultman and S.A. Barnett, Appl. Phys. Lett. 57, 2654(1990).
54. M. Shinn, L. Hultman and S.A. Barnett, J. Mater. Res. 7, 901(1992).
55. X. Chu, M.S. Wong, W.D. Sproul and S.A. Barnett in Ref. [51].
56. J.E. Krzanowski, MRS Symp. Proc. 239, 509(1992).

57. A.K. Head, Phil. Mag. 44, 92(1953).
58. E.S. Pacheco and T. Mura, J. Mech. Phys. Solid. 17, 163(1969).
59. W.D. Nix, Met. Trans. 20A, 2217(1989).
60. For review articles, see for example P.S. Ho and T. Kwok, Rep. Prof. Phys. 52, 301(1989);
    A. Scorzoni, B. Neri, C. Caprile and F. Fantini, Mater. Sci. Rep. 7, 143(1991).
61. R.G. Filippi, G.A. Biery and M. Wood, MRS Symp. Proc. 309, 141(1993).
62. F. Xu, J.F. Shepard, Jr., T. Su and S. Trolier-McKinstry, MRS Symp. Proc. 493, (1998).
63. G. Teowee, K.C. McCarthy, F.S. McCarthy, D. Davis, J.T. Dawley, R. Radspinner,
    B.J.J. Zelinski and D.R. Uhlmann, ibid.

Appendix

A1. Flow and fracture of thin films.

A1.1. Yield strength.

With the onset of an awareness of the nanodimensional world into our collective consciousness many additional studies have been made with layer and/or film thickness dimensions in the nano realm in the decade previous to this writing (2005). Figure A4.1 reveals results obtained on free films (in the absence of substrates) as

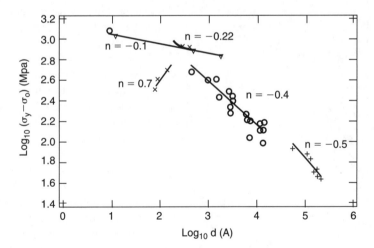

Figure A4.1. Comparison of the yield stress of Ag/Cu multilayers: O (Huang and Spaepen[A1]); ∇ nanoindentation (Verdier et al.[A2]); + coarse-grained Cu (Choksi et al.[A3]); nanocrystalline Cu x (Choksi et al.[A4]); and ◆ (Sun et al.[A5]). (Reproduced from Huang et al., Int. J. Fracture, 119/120, 359(2003).)

well as yield strengths derived from indentation observations.[A1] The main point arising from these data is that the exponent in the Hall–Petch relation changes from −0.5 to −0.1 below about a layer thickness of about 10 nm. The values indicating a positive value to this slope are believed to be an artifact due to the closing up of pores in the samples. Figure A4.2 provides additional data indicating the saturation in hardness (and yield strength) with decreasing layer thickness.[A2] One of these set of data indicates a maximum in the dependence on layer thickness which will be discussed later. For all the multilayer data in these figures the significance of the absolute values of their hardness and yield strength in comparison to theoretically predicted values is questionable because the internal stresses in the layers due to coherency, interfacial stresses and, perhaps, other sources are not taken into account in these theories. That such internal stresses exist in multilayers has been demonstrated.[A3] Figure A4.3 shows values of internal stresses measured in Ag–Ni multilayers. As shown, the internal stress in the very thin Ni layers has the same magnitude as the yield strength values (hardness/3) in Figure A4.2. As already noted most attempts to provide an explanation for data such as shown in Figure A4.1 and A4.2 do not consider the effect of the internal stresses in multilayers on the hardness or yield strength. Nevertheless, empirically it is apparent that both hardness and yield strength increase with decrease in the layer thickness.

Most of the multilayers are deposited at room temperature or below using either sputtering or evaporation as the source of the depositing atoms. It is gratifying that the density of these films is reported to be close to that of the bulk, a result that this writer would not expect since room temperature is usually below the critical temperature separating deposition of voided films from void-free films.[A4] However, at least one component of these two-component multilayer configurations with corresponding data in Figures A4.1 and A4.2 has a surface diffusion length at room temperature larger than the layer thickness (and lateral grain dimension) for values of the latter less than about 100 nm so that its diffusion would act to fill any intergranular voids that

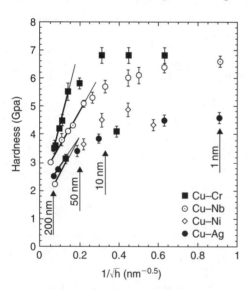

Figure A4.2. Hardness versus inverse square root of thickness. (Reproduced from A. Misra, J.P. Hirth and H. Kung, Phil. Mag. A82, 2935(2002).)

form during the deposition of the other component. No evidence for such filling has been reported to the knowledge of this writer. Detection of such an effect is difficult, however.

With respect to the internal stresses developed in multilayers the writer would like to remark that in the first book to this series he noted that the magnitudes of the internal stresses developed in thin films deposited on substrates at room temperature invariably corresponded to the yield strengths. We may note that the physical origin of the tensile internal stresses in thin films on substrates also acts in multilayers in addition to the interfacial sources of such stresses. This physical origin is the cohesive attraction of surfaces in close contact, these surfaces being the surfaces of adjacent grains that are transverse to the substrate surface and which are in the process of growth during deposition. This cohesive attraction tends to decrease the area of the layer, a tendency that is opposed by the coherence of the layer to the one below it and so on to the substrate. In this way a biaxial tensile stress is developed in the growing layer. Although this phenomenon occurs in both component layers, the magnitude of the tensile stress developed in a growing layer is not the same in the different components since this magnitude is limited by the yield strength which differs between component materials. Thus, the tensile stress in the component layer of lower yield strength will change to a compressive stress upon growth upon it of a tensile stress layer of much higher yield strength. The multilayer film stress situation is complicated relative to the single layer film by the effect of interfacial stress as well.

The softening indicated in Figure A4.2 by the open diamond shaped points has been found for other multilayer systems.[A5] This writer is not able to conclude whether this feature is an artifact due to the presence of voids in the samples or is a phenomenon characteristic of a dense multilayer system. There are many explanations in the

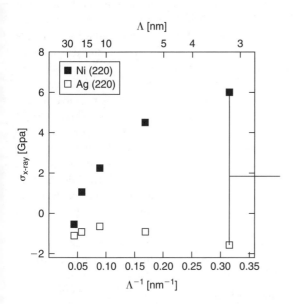

Figure A4.3. Internal stresses in Ag–Ni multilayer films versus inverse periodicity. (From D.D. Fong, **Ph.D. Thesis, Applied Physics**, Harvard University, 2001 with permission.)

literature for this effect, sometimes called the inverse Hall–Petch effect. An evaluation of the reality of this effect implies that it is real.[A6]

A1.2. Fracture.

The significant new structures of thin films that have appeared in the previous decade are related to nanoscale aspects, to biologically inspired composite structures, and to free-standing films as in MEMS. As expected, the reduction of grain size to nanoscale dimensions by reducing the scale of flaws in the films results in an increase in the fracture strength.[A7] However, when the scale of the grain size becomes less than that of the surface roughness the latter controls the statistics of fracture.[A8]

The last decade has also seen an increase in the number of studies of fatigue behavior of thin films as compared to the previous era. One result is shown in Figure A4.4 where the fatigue limit of Ag thin films (stress amplitude below which no fatigue failure occurs in $3.8 \cdot 10^6$ cycles of stress) is plotted versus the inverse square root of the film thickness.[A9] As shown, the data obey a Hall–Petch relation, as might be expected since fatigue failure in a ductile metal requires plastic strain.

Given the need for plastic deformation it is surprising that silicon thin films exhibit fatigue failure.[A10] However, this mode of failure has been found to involve the oxide film covering the Si film and to be induced by a reaction-layer fatigue process that involves the sequential, stress-assisted oxidation and stress-corrosion cracking of the native oxide.[A11] When the Si surface is protected from oxidation the fatigue type failure does not occur.[A12]

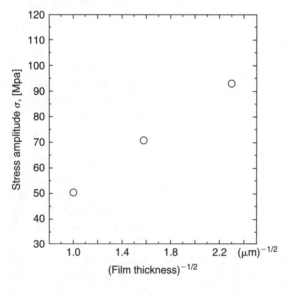

Figure A4.4. Stress-amplitude in fatigue testing below which failure does not occur in $3.8 \cdot 10^6$ cycles. (Data taken from R. Schwaiger, **Dissertation**, University of Stuttgart, December 2001.)

References to Appendix

A1. H. Huang et al., Int. J. Fracture, <u>119/120</u>, 359(2003).

A2. A. Misra, J.P. Hirth and H. Kung, Phil. Mag. A<u>82</u>, 2935(2002).

A3. D.D. Fong, **Ph.D. Thesis, Applied Physics**, Harvard University, 2001.

A4. E.S. Machlin, in **Materials Science in Microelectronics I, The Relationships between Thin Film Processing and Structure**, Elsevier, 2005, Chapter 2.

A5. A. Misra, M. Verdier et al., Scripta Mater. <u>39</u>, 555(1998); C. Kim, S.B. Qadri et al., Thin Solid Film. <u>240</u>, 52(1994); S. Tixier, P. Boni, H. Van Swygenhoven, Thin Solid Film. <u>342</u>, 188(1999); J.R. Weertmann et al., MRS Bull. <u>24</u>, 44(1999).

A6. C.C. Koch and J. Narayan, MRS Symp. Proc. <u>634</u>, B5.1.1.(2001).

A7. B. Peng and H.D. Espinosa, Proceedings of the IMECE 2004. (ASME International Mechanical Engineering Congress 2004).

A8. H.D. Espinosa et al., J. Appl. Phys. <u>94</u>, 6076(2003).

A9. R. Schwaiger, **Dissertation**, University of Stuttgart, December 2001.

A10. C.L. Muhlstein, S.B. Brown and R.O. Ritchie, MRS Symp. Proc. <u>657</u>, EE5.8.1(2001); J. Microelectromech. Syst. <u>10</u>, 593(2001).

A11. P. Shrotriya et al., MRS Symp. Proc. <u>687</u>, B2.3.1(2002).

A12. C.L. Muhlstein, E.A. Stach and R.O. Ritchie, ibid., B10.7.1.

Problems

1. How would you increase the yield strength of a thin film?
2. If it was necessary to prevent a brittle film, such as a polycrystalline ferro-electric DRAM capacitor, from developing cracks across its cross-section what steps would you take to assure this result?
3. What steps could you take to prevent delamination of an oxide film from an oxide substrate?
4. Given a ductile metal film that does not bond strongly to a substrate. How would you alter the structure at the interface to improve the delamination strength?
5. Voids are found in a polycrystalline thin film after an annealing cycle. How would you alter the structure of the thin film to reduce the propensity for thermal stress voiding?
6. Why should diminishing the distance between mass sources and conjugate sinks during electromigration act to decrease the rate of mass flow?
7. Why would increasing the yield strength of the conductor line act to diminish the rate of mass flow during electromigration?
8. Why should a difference in the number of grain boundaries on the upstream side of a conductor line cross-section as compared to the downstream side lead to a divergence in the mass flow during electromigration?

9. When would you expect that a fine-grained film would yield a smaller contact fracture fragment than a large-grained film?
10. What relation normally determines the contact area between contacting materials?
11. Suppose two materials were in contact, what property would determine the fraction of the contacts that would fracture at other than the interface?
12. How does hardness exert its effect on friction and wear?

CHAPTER V

Mass Transport Properties

Because of the fine grain size in most as-deposited polycrystalline films, grain boundaries control the observed mass transport, at least at temperatures less than one-half of the absolute melting temperature. Thus, a study of the effect of structure on grain boundary diffusion is one objective of this chapter. However, mass transport also occurs in monocrystalline films where the point defect structure controls the rate of transport through the lattice. Surface diffusion is also of interest primarily in deposition processes, but also in applications as well. We will examine the effect of structure on all three modes of transport. Further, we will discuss examples where these various types of transport affect properties. Some thin films are amorphous and mass transport in some of them has practical import. Consequently, we will devote some discussion to this topic. Diffusion barriers are used to prevent diffusion between layers. Hence, we will examine what aspects of structure affect the ability of diffusion barrier materials to prevent such atom transport and to prevent chemical reactions with the adjacent layers. The effect of pre-existing structure on the course of reactions between layers will also be considered. Finally, the effect of defects introduced via radiation on the rates of reactions will be discussed.

1. Grain boundary diffusion and its effect on properties.

1.1. Grain boundary diffusion.

Five parameters are necessary to define grain boundaries: the orientations of the bounding grains (2), the orientation of the boundary relative to one of the grains (3). It is not surprising therefore that only a small fraction of this parameter space has been explored. Nevertheless, we can characterize grain boundaries grossly as either large angle or small angle boundaries. The latter type range from pure tilt to pure twist containing only edge and screw dislocations, respectively, with mixtures of both dislocations in mixed tilt/twist boundaries. The transition between small and large angle boundaries occurs when the cores of the dislocations providing the tilt or twist motions producing the relative orientations of the bounding grains overlap. Sufficient experimentation has been performed that the following generalization can be made: self-diffusivity along screw dislocations is much smaller than along edge dislocations. Indeed, for the specific case of <111> twist boundaries in Ni and Ag the ratio of the activation energy to that for bulk diffusion is 0.7, whereas for <100> tilt boundaries this ratio is 0.4 and for <211> tilt boundaries the ratio is 0.6, again in both Ni and Ag. It is also known that the self-diffusivity along dissociated dislocations is smaller than along undissociated ones. These facts correlate with our intuition based evaluation of the average atomic volume within a given radius about the various dislocation cores. The higher activation energy and lower self-diffusivity is associated with tighter packing of the atoms in the dislocation cores of the extended dislocations. As may be expected the grain boundary self-diffusivity parallel to dislocation cores is higher than in the direction perpendicular to them, i.e. the self-diffusivity is higher parallel to the tilt axis of tilt boundaries than perpendicular to it. Twist boundaries do not exhibit such transport anisotropy.

There are differences in self-diffusivity along high angle grain boundaries which relate to a difference in grain boundary structure. This difference in grain boundary structure is revealed indirectly in a difference in the grain boundary energy. A polar plot of grain boundary energy (a Wulff plot) reveals energy minima at cusps that occur at certain high angle misorientations. Such a plot is shown in Figure 5.1. The grain boundaries corresponding to the cusps are denoted as special boundaries. These special grain boundaries exhibit a tighter packing than those away from the cusps. Thus, it would be expected that the grain boundary self-diffusivity for the special boundaries would be less than for those boundaries not at cusp misorientations. This expectation is mirrored in a relationship proposed by Borisov[1] between grain boundary energy and the ratio of grain boundary to lattice self-diffusivity. Some evidence for such behavior has been obtained by Gupta.[2]

Grain boundary energy will be lowered by Gibbs segregation of solute to the grain boundaries when the composition of the parent phase is within its region of thermodynamic stability, i.e. segregation at grain boundaries of solutes within the spinodal compositions will increase the grain boundary energy. (See Thermodynamics in the bibliography.) Segregation decreases the grain boundary energy when there is an increase in the total bonding strength of the atoms at the grain boundary per unit area of boundary, i.e. when the energy of mixing is negative. Thus, we may expect that substitutional solute atoms that segregate to the grain boundaries, that have higher cohesive strengths than that of the solvent, and that have about the same atom radius as the solvent atom will act to decrease the grain boundary energy. In this case, there should be no change in the number of bonds per unit area of boundary and in the average excess specific volume, although there will be an increase in the number of stronger bonds there. Thus, solvent atom diffusion (self-diffusion) along such a boundary will require that stronger solute–solvent atom bonds be broken than for the case of a pure host and, hence, self-diffusivity in the alloy should be lower than in the pure host. This expectation is supported by the observed effect of Ta as a solute in decreasing the self-diffusivity of Au in its high angle grain boundaries.[3] All the conditional statements above

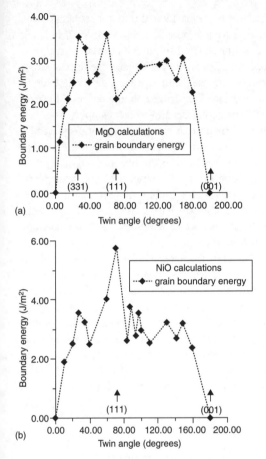

Figure 5.1. Grain boundary energies as a function of twice the tilt angle (twin angle) about $[1\bar{1}0]$ tilt axis for (a) MgO and (b) NiO. from Phys. Rev., B60, 2740(1999) with permission. Copyright 1999 by the American Physical Society.

should be noted for it is possible that violation of one of them will yield a case where decrease in the excess grain boundary energy is not accompanied by a decrease in the self-diffusivity along grain boundaries.

　　Another case where segregation that decreases the grain boundary energy may lower the self-diffusivity is that produced when the solute atom is of interstitial type, segregates at the boundary without changing the spacing between atoms there (the hole it occupies at the boundary has a volume about equal to its atomic volume) and thereby acts solely to increase the number of bonds to solvent atoms at the boundary. Boron may act in such a manner in Ni based hosts.[4]

　　The effect of segregation of atoms having higher cohesive energy than the host to grain boundaries of the latter on the self-diffusivity may be evaluated by a comparison of the diffusivity of the solvent component in a pure host grain boundary with its diffusivity at an interphase interface between the pure host and pure solute phases. For example, the activation energy for self-diffusion of Ag along its high-angle grain boundaries ranges from about 15 kcal/mol to 26.4 kcal/mole, whereas along an interface between pure Ag and pure Fe it equals 45 kcal/mole and between a Ag rich solid solution and a Cu rich solid solution it equals 65 kcal/mole. Other factors in addition to the bonding effect must act to increase the activation energy of Ag in the interphase interfaces because the increase shown is much higher than would be expected based on the increase in average bonding energy. However, the data certainly do not disprove the expected effect and, in fact, support it.

In thin films, grain boundary diffusion provides the mass transport at the low temperatures involved in the phenomena of electromigration, in the production by interdiffusion of most silicide phases, and in the use and degradation of certain multilayer combinations. Actually, the total fraction of mass transported per unit area by the grain boundary decreases with increasing temperature. Below about $0.5 \cdot T_m$, where the latter parameter is the absolute melting temperature, grain boundaries transport the major fraction, whereas above this value mass transport via the lattice accounts for the major fraction.

Unfortunately, not enough data exist to enable a prospective user to obtain needed information from a data bank, such as Ref. [5]. However, certain generalizations can be used as a basis for estimating grain boundary self-diffusivities. In particular, in fcc metals the activation energy for self-diffusion along high angle grain boundaries is estimated to equal $17 \cdot T_m$ in cal/mole, with a pre-exponential factor equal to $0.3 \, cm^2/sec$. For bcc metals there are fewer data on which to base a similar generalization, but for the two metals for which data exist the activation energy ranges between $20 \cdot T_m$ and $25 \cdot T_m$. As a rough guide it is apparent that the activation energy for self-diffusion along a given grain boundary structure increases with the melting point of the material.

1.2. Effect of grain boundary diffusion on properties.

1.2.1. Conductivity during electromigration.

As mentioned in a previous section, most of the mass transport induced by the passage of electric current along interconnect lines (i.e. the phenomenon of electromigration) occurs along grain boundaries. The divergence of this mass transport leads to the development of voids and hillocks in the interconnect lines and ultimately to failure of the interconnect line, either by the formation of a complete open to cause a cessation of current, or by hillock connection to an adjacent conductor to form a short-circuit. Obviously, as voids develop the resistance of the interconnect increases. Figure 5.2 shows the dependence of the resistance on the time the interconnect line is subject to current. The initial decrease in resistance evident in Figure 5.2 is a consequence of the removal of Cu from solid solution via precipitation. The subsequent increase in resistance is due to the generation of

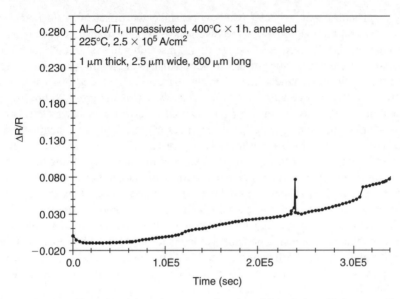

Figure 5.2. Normalized change in resistance as a function of time during which current flows along interconnect line. From C.-C. Lee, D. Eng. Sci. Thesis, Columbia University, 1991.

vacancies and voids. The spike in resistance shown in Figure 5.2 at $2.4 \cdot 10^5$ is due to the formation of an open in the conductor layer of Al–Cu, which then closes upon the thermal expansion generated by the local heating when the current is shunted through the adjacent Al_3Ti shunt layer. In the absence of an electrically conducting shunt layer the Al–Cu conductor layer would have developed an open circuit at the time corresponding to the first resistance spike. The effect of the shunt layers is to increase the time to failure by a factor of about 2 for a single shunt layer and 4 for two shunt layers for the experimental conditions corresponding to the result shown in Figure 5.2.[10]

Possible modes of forestalling such electromigration induced failures are: (1) decrease the self-diffusivity along the grain boundaries; (2) decrease the magnitude of the divergences in mass transport; (3) increase the counter gradient of the stress opposing the mass transport.

The use of copper as a solute in Al interconnect lines has resulted in a decrease in the self-diffusivity of Al along grain boundaries in such interconnect lines as revealed by an increase in the activation energy associated with electromigration induced failure.[6] Similar effects of solute segregation at grain boundaries on the self-diffusivity along these boundaries may be expected for appropriate solutes as described in section 1.1 above. Generally, the stronger the tendency for solutes to segregate at grain boundaries the slower is diffusion along the boundaries. A review of the tendency for grain boundary segregation is provided by Hondros and Seah.[7]

There are no reliable results concerning the effect of grain boundary parameters on the magnitude of the divergences in mass transport developed in interconnect lines during electromigration. This ambiguity is a consequence of ignorance of the diffusivities of the grain boundaries at the grain boundary junctions present in interconnect lines. For, the extreme case of a bamboo structure, where some grain boundaries extend completely across the cross-section of the interconnect line, the presence of sites where divergence in mass flow occur becomes obvious. Similarly, larger divergences may be expected when the grain size is sufficiently large so that there is a large variation in the number of grain boundaries transporting mass along the length of the interconnect line than when the grain size is finer and the percent variation in this number is smaller. In the extreme and unlikely case where no divergences in mass occur along the length of an interconnect line then mass would build up at one end of this line and be depleted at the other end of this line. In principle this would occur in a monocrystalline interconnect line and, perhaps, could be accommodated by design without bringing about failure of the interconnection. One aspect of grain boundary structure relating to texture and its effect on developing mass divergence has been explored. In particular, grain boundaries corresponding to an ideal {111} texture in Al would be tilt boundaries with the tilt axis normal to the film plane. In this situation diffusion in the film plane in such boundaries would be slower than in the direction normal to the film plane. Also, and more important, the boundary between a grain having the ideal {111} texture and one with an orientation rotated away from this texture should have a much higher self-diffusivity along the film plane than for the boundary between two ideally oriented {111} grains, except for the special boundaries at cusp orientations. Thus, local deviation from the ideal texture should affect the incidence of local mass divergence. Just this result was observed in that mass divergences, in the form of voids or hillocks, were found to be associated mainly with grains having orientations deviating significantly from the ideal {111} orientation, in films exhibiting a predominant {111} texture.[39]

The counter force opposing mass transport due to electromigration is that developed by the stresses induced at the mass divergences. Mass transport induces tensile stress near the source and compressive stress near the corresponding sink. The gradient in this stress opposes mass diffusion due to electromigration.[8] The use of capping and passivation layers on interconnect lines acts to increase the maximum compressive stress developed in the interconnect lines, without affecting the tensile stress which tends to remain nearly nil due to void formation, and for the ideal case this compressive stress is sufficient to make the net mass flow cease.[9,10] Much work needs to be done to discover the grain boundary structure (crystalline texture, in-plane orientation, grain size) that optimizes the resistance to electromigration in interconnect lines and to develop capping layer materials of higher strength. Some work has begun with the objective of increasing the yield strength of the interconnect lines.[11] One way this may be achieved is suggested by the effect of film thickness on yield strength described in the previous chapter, i.e. use multilayer interconnect lines in which the conducting layers are separated by much stronger, higher modulus capping layers.

As line widths get smaller the percentage of bamboo-like grains in an interconnect line increases. However, not all grains are bamboo-like and mass-divergences will occur at the intersection of a multigrain section of the inter-connection with a bamboo-like grain. This is an aspect of microstructure where control of the maximum distance between bamboo-like grains along the intercon-nection can be crucial to the concomitant control of the time to the first open or electromigration failure of the interconnection. If this maximum distance between successive bamboo-like grains along the interconnection is larger than the critical source-sink distance (defined in the previous chapter), then the counter force due to the gradient of the stress on the electromigration flux of mass will be insuffi-cient to cut-off this flux, and the presence of such bamboo-like grains may decrease the time to failure as compared to a bamboo-like-grain-free interconnec-tion. However, should this maximum distance between bamboo-like grains be less than the critical source-sink distance, then the flux of matter during current flow would be blocked and the time to failure would increase enormously. Unfortu-nately, the critical source-sink distance has a value somewhere between 10 and 50 μm[10] for Al based interconnections and the statistics of achieving the complete absence of concomitant source-sink distances larger than this critical maximum value, especially along a long interconnection, is discouraging.

Thus, electromigration along grain boundaries has until now exerted a detrimental effect on the lifetime and resistance of interconnect lines. An under-standing of the effect of structure on this phenomenon has led to improvements in electromigration resistance with still further improvements in the offing.

1.2.2. Effect of grain boundary diffusion on reaction temperature.

The production of integrated circuits often involves the formation, by reaction between two adjacent layers, of a phase that is present in the phase dia-gram of the binary system formed by the two component layer materials. For example, silicide electrodes are sometimes formed by this procedure by the diffu-sive reaction between silicon and a metal layer. Also, aluminide capping layers are formed on aluminum interconnect lines by such a diffusive reaction. There are many other examples that can be cited where such a procedure produces a useful product in microelectronics. *Grain boundaries are important in such reactions because they provide the means for reducing the temperature at which the desired reaction can occur.* (The trend for miniaturization drives down the maximum tem-peratures that integrated circuits can be allowed to experience.)

How do grain boundaries bring about a lowering in the reaction tempera-ture? First, by acting as preferred nucleation sites for the reaction product phase.

Second, and most important, by acting as short-circuit, rapid diffusion paths for the most mobile reactant. In this way the latter can come into contact with the other reactant and thereby allow the product phase to grow at a lower temperature than would be required for the case where diffusion is through the lattice of the product phase.

From the foregoing, it is obvious that the grain boundary diffusion paths involved in the reaction between the reactant layers are in the product phase. Is there any relation between these grain boundaries and the grain boundaries in the reactant phases? The grain boundaries of one of the reactant phases, or the interface between the reactant phases, or both will provide nucleation sites for the product phase. Whenever, the grain boundaries of a reactant phase act to promote product phase nucleation then the *grain size* of the product phase will be on the order of that for this reactant phase, i.e. product phase grains nucleated at reactant phase grain boundaries will grow laterally until they impinge. Thus, if the grain size of the reactant phase is small (as it is in most deposited films), then it will be similarly small in the product phase, and the reaction temperature will be lower in thin films than between bulk reactants having much larger grain size. Indeed, this is a result often cited in the literature relating to silicide formation in thin films. Thus, the grain boundary structure of one of the reactant phases often does affect the kinetics between the reactants in thin films simply by controlling the grain size in the resultant product phase.

A recent paper by Gas and d'Heurle[12] emphasizes the important role of grain boundary diffusion in silicides. It appears that the ratio of the grain boundary diffusivity to that for the lattice is higher in silicides than in metals so that grain boundary diffusion is the controlling mechanism in silicides to even higher growth temperatures than in metals.

There are many other examples where the influence of grain boundaries in the formation of desired or undesired microstructures via diffusive reactions in thin films affect properties. For example, a fine grained film of permalloy, which normally would not exhibit a giant magnetoresistance, has been induced to have this property by the simple means of diffusing Ag into its grain boundaries so as to produce a thin film of Ag separating the NiFe grains.[13] The number of potential undesired reactions between adjacent layers, or between layers separated by another layer, that are mediated by grain boundary diffusion are too numerous to list. These undesired reactions are the *raison d'être* for diffusion barriers.

1.2.3. Effect of grain boundary diffusion on hillock formation.

Usually, the formation and growth of hillocks on surfaces of thin films is induced by the presence of a compressive stress in the plane of the film. Transport of mass from within the film to the film surface, where this mass forms a hillock that is stress-free, relieves the compressive stress within the film. The fastest mode of mass transport of atoms within the film to the film surface is via grain boundary diffusion. Thus, hillocks formed in this manner will be centered on the intersections of grain boundaries with the surface. This type of hillock formation is found in many cases. For hillocks that are formed by grain boundary transport of mass to them, the rate of hillock formation can be decreased by concomitant decrease in the rate of grain boundary mass transport. One technique of reducing self-diffusion along grain boundaries we have already met is that of making use of an appropriate grain boundary segregant. It is believed that Cu[3] and various rare-earth elements[21] act in just this manner in Al interconnects. Unfortunately, many of the experiments conducted to evaluate effects of grain boundary diffusion on hillock formation are not definitive in that the possible role of independent factors on the grain boundary mass transport were not tested. For example, in one series of experiments where it was shown that implants of N, O and Ar with doses of about $10^{16}/cm^2$ and implant depths from 60 to 400 Å markedly reduced hillock density in annealed Al films containing 1 wt% Si, no observations were made to determine whether or not there was a recrystallized layer in the implant region.[22] Separate studies (see *Volume I of Materials Science in Microelectronics*, p. 76) have shown that such energetic bombardment does act to recrystallize the surface layer. Compressive stress need not be the only driving force for hillock formation. For example, it has been found that the inward diffusion of P as a dopant along grain boundaries in poly-Si acts to induce the reverse grain boundary self-diffusion of Si along its boundaries and thereby produce hillocks.[23]

2. Lattice diffusion and its effect on properties.

Lattice diffusion in crystalline materials is affected by the point defect structure, which itself depends on the crystal structure and type of bonding. This subject has been treated in textbooks[14] and the same principles apply to thin films as to bulk materials. Thus, the discussion to follow will be brief and will focus on the effects of parameters likely to be found significant only in thin films.

Thin films can support high biaxial stress and many usually do. Thus, an effect on the equilibrium number of point defects due to such stress is to be expected. The question is whether or not this effect significantly affects lattice diffusion in thin films. The answer is that the quantitative effect needs to be evaluated for each specific case. For example, for aluminum, where lattice diffusion occurs primarily by vacancy–atom interchange, where the work to form a vacancy equals about 0.66 eV and the associated volume of formation equals about 0.35 of an atomic volume,[15] and in which the biaxial stress can reach on the order of $10^9 \, Pa$[16], the work to form a vacancy can be changed by about 5.4%. Further, the effect of stress on the enthalpy of vacancy migration in Al is smaller due to a smaller activation volume of migration and an enthalpy of vacancy migration nearly equal to that for vacancy formation. Thus, the effect of stress on lattice diffusion in Al thin films is negligible. On the other hand, semiconductors, through the effect of stress on the band gap, may be more sensitive than metals to the effect of stress on lattice diffusion. One such effect of stress has been suggested in the doping of silicon by arsenic.[37]

The other factor affecting lattice diffusion that might be accentuated in thin films is the non-equilibrium concentration of point defects. The vacancy concentration is higher than the equilibrium concentration in films deposited using non-energetic particles only when the absolute substrate temperature is less than about $0.3 \cdot T_m$ (see *Volume I of Materials Science in Microelectronics*, p. 39). For films deposited using energetic particles impinging upon the film during deposition there may well be high non-equilibrium concentrations of vacancies and other defects. It is not possible to provide a quantitative estimate of this effect on lattice diffusion because the defect concentrations introduced into films during energetic particle deposition can vary orders of magnitude depending upon the deposition conditions, and energy and fluence of impinging particles. For some films, such as silicon, oxidation can introduce non-equilibrium concentrations of defects, such as interstitials, which may act to enhance the local self-diffusivity and also form dislocation loops upon precipitating out when the concentration becomes sufficient to nucleate these loops. Further, the growth on silicon of films due to interaction with a deposited layer can act to change the defect concentration in the subinterface region and thereby alter the local lattice diffusivity. We will describe examples of both of the latter phenomena in the next subsection.

The activation energy for lattice self-diffusion in fcc metals is about $34 \cdot T_m$ cal/mole, where T_m is the absolute melting temperature. Studies of diffusion of metal species in amorphous alloys indicate that the activation energy is about $60 \cdot T_g$ in cal/mole, where T_g is the glass temperature with a preexponential factor of about 10^{-4} cm^2/sec. Thus, diffusion in amorphous alloys is appreciably slower than in crystalline metals. This fact has led to the use of amorphous materials as diffusion barrier layers. We will discuss this subject later in this chapter under the section devoted to diffusion barriers. Lattice diffusion in semiconductors, and especially compound semiconductors, as well as in ionic compounds, is more complicated than in metals. The mechanism of lattice diffusion in many semiconductors is still unsettled. A discussion of these subjects is outside the scope of this book and the interested reader is referred to publications on this topic. See bibliography.

2.1. Reaction layer induced lattice defects in silicon.

It is well known that oxidation of silicon that does not proceed via a strictly planar interface acts to pump interstitial Si into the Si lattice beneath the interface with the oxide.[38] This process occurs because the volume occupied per Si atom is larger for the oxide than for the elemental Si. These interstitials can condense to form stacking faults and prior to condensing act to enhance the lattice diffusivity. Indeed, this consequence of Si oxidation is now used routinely to getter impurities from Si wafers. Analogous to the effect of oxidation in the production of interstitials in Si it has been found that silicidation also can change the point defect concentration adjacent to the interface between the silicide and the Si substrate. For example, the formation and annealing in the temperature range 800–890 °C of a $TiSi_2$ film on Si caused a vacancy supersaturation and an interstitial undersaturation in the underlying Si.[24] Such point defect anomalies have concomitant anomalies in the lattice diffusion of dopant atoms and their profiles, a result that becomes significant as miniaturization proceeds.

3. Surface diffusion and its effect on properties.

Surface diffusion may act to affect surface morphology, such as in the growth of hillocks,[17] the smoothening of a rough surface,[18] the development of grain boundary grooves, etc. It may also contribute to a shortening of the lifetime of interconnect lines under the influence of the electron wind.[19] Surface diffusion during deposition also controls a variety of structural aspects, which in turn affect properties. For example, insufficient surface diffusion during deposition at substrate temperatures less than T_1 results in films having less than the ideal bulk density. Also, inadequate surface diffusion during deposition onto monocrystalline substrates can result, as it does in the case of silicon, in the formation of an amorphous film. Hence, there are schemes for enhancing surface diffusion. For example, surface active monolayers on surfaces are used to allow for lateral diffusion of depositing atoms prior to their attachment to the film surface.[20]

One may well wonder how it is possible for surface diffusion to roughen a surface by producing hillocks and also to smoothen it. The explanation of this paradox is that what occurs depends upon the driving force for the surface diffusion and not upon the latter alone. When the removal of atoms from the surface to produce hillocks decreases the free energy then it may occur. If the strain energy per atom in the original surface exceeds the increment in surface energy per hillock atom due to the hillock, then the free energy is reduced by this mass transfer, providing that the hillock has a small enough radius so that there is no strain energy in the hillock due to the constraint on it exerted by the film. Similarly, smoothening of a rough surface of a film may occur, if such smoothening does not lead to an increase in the strain energy experienced by the transferred atoms, because the total surface energy (surface area) is reduced by this process.

A possible method of enhancing surface and interface diffusion in semiconductor films will be discussed in the last section of this chapter.

4. Diffusion barriers.

There are many examples among thin film products where the reaction between adjacent layers results in a deterioration of properties. Diffusion barrier layers were developed to hinder such interaction. The principles governing diffusion barriers are simple. We have already noted that the activation energy for diffusion in metals scales with the absolute melting temperature. Thus, the higher is the melting point of a metal the higher will be the activation energy and the lower will be the diffusivity (self or solute) at a given use temperature. Thus, most diffusion barriers are based on the use of materials that have a high melting point. Because grain boundaries are the paths of fastest diffusion, it is desirable in diffusion barrier applications to use a material without grain boundaries, such as an amorphous material. Finally, if the use of amorphous materials is not possible for some reason then the grain boundaries of the diffusion barrier material should either be low energy boundaries, if possible, or the boundaries should be "stuffed" with segregant atoms that lower self and solute diffusivity. In addition to all the foregoing principles there is another that applies to all diffusion barrier materials. Namely, that the diffusion barrier material should not react with, nor should any components of the diffusion barrier material diffuse into, its contacting layers. Despite this knowledge of general principles it is usually necessary to test candidate diffusion barriers. A summary of the knowledge we have about diffusion barriers may be found in Ref. [25].

Very often, because diffusion barriers are not elements, it becomes necessary to make use of additional diffusion barriers in series, each one designed to act as a barrier for a component of the complex assembly of substrate/diffusion barrier/device. For example, amorphous TiB_2 acts as an efficient diffusion barrier to copper for temperature anneals up to 750 °C for 30 min.[26] However, B from TiB_2 diffuses into silicon.[26,27] But, boron does not diffuse through $TiSi_2$ up to 900 °C.[28] On the basis of these facts, Sade and Pelleg[29] examined the possibility of using $TiSi_2/TiB_2$ as a series diffusion barrier between silicon and copper.

The microstructure of the diffusion barrier can have an effect on its performance as a barrier. For example, a diffusion barrier layer with columnar grains in which the diffusant can be transported along the columnar boundaries is likely to be a less resistant barrier than one in which the grains are equiaxed and the crystallographic texture random.[30] Also, when the grains in the diffusion barrier are columnar it is useful to consider a series arrangement to break up the straight diffusion path. For example, TiN is usually deposited with a columnar microstructure. By combining it in series with another layer with which it is inactive chemically and which has a non-columnar grain structure or is amorphous, the effective diffusion barrier may be enhanced.[31]

5. The effect of structure on reactions in thin films.

This is a subject that has been studied extensively in bulk phases and one aspect of this subject forms the field of heterogeneous nucleation. We have already referred to the effect of grain boundaries acting as heterogeneous nucleation sites and as short-circuit paths for the diffusion of a reactant involved in a reaction in a thin film. Usually, kinetic considerations govern the speed with which reactions take place in thin films.[32] However, heterogeneous nucleation can govern the density of short-circuit diffusion paths produced in a product phase and in this way indirectly affect the speed with which a product phase grows, i.e. the time of nucleation of the product phase is not usually the time limiting the appearance of the product phase in thin films except under conditions near equilibrium when the free energy driving the transformation that appears in the denominator of the work of nucleation* approaches zero and, hence, the work of nucleation becomes very large.

Although nucleation is not the time-limiting process in most thin film reactions (i.e. the rate of growth of nuclei into product phases is the usual rate limiting process) it is almost certainly true that the product phase is heterogeneously nucleated in all these cases. When the supply of heterogeneous nucleation sites becomes negligible then the nucleation event can become the time controlling process in the reaction. Indeed, this situation can occur when the line-width becomes less than the grain size in processes of siliciding of metal lines. Just this case has been observed in the C49 to C54 phase transformation in $TiSi_2$.[34] Two processes have been used to increase the density of heterogeneous nucleation sites for this transformation. One successful method was to pre-amorphize the Si substrate prior to depositing the Ti and to use high temperature sputtering as the deposition mode.[35] Another method used a controlled dose of molybdenum $(5 \cdot 10^{13} \text{atoms/cm}^2)$ as an implant in the silicon prior to titanium deposition.[36] For a reason yet unknown the C49 to C54 transformation temperature was lowered and the grain size of the C54 silicide was refined. In some way the molybdenum facilitated the nucleation of the latter phase. Further studies of the latter effect involved the deposition of Ti alloy films containing Mo, Ta, or Nb as alloying agents in amounts up to 20 at.%.[40] It was speculated that a C40 refractory metal silicide phase on the surface of the Ti alloy film acts as a template for the C54 $TiSi_2$ nucleation, i.e. by decreasing the surface energy barrier to nucleation.

Examples of the effect of defect structure in the parent phase upon the kinetics of reactions and the microstructure of the product in thin films abound.

*For an isolated nucleus in a parent phase, the work of homogeneous nucleation is given by $\Delta G^* = \beta \sigma^3/(\Delta g)^2$, where β is a numerical factor depending upon on shape of the nucleus, σ is the specific energy of nucleus/host interface and Δg is the specific free energy difference between parent phase and nucleus – the driving free energy.

These effects are accentuated at low temperatures when the difference in the diffusivity of the fastest diffusing species through the lattice and the defects is large, but the lattice diffusivity is still sufficient to allow for the lateral transfer of the diffusing species to form layers of reaction product adjacent to the defects. Extended defects, such as stacking faults, anti-phase boundaries, low angle boundaries, double position boundaries and grain boundaries, are the sites of preferential diffusion and the regions at which the reactions first occur. In thin films where the grain size is small, this process can lead to the complete transformation of the film from the parent structure to the product structure with the product phase *inheriting the grain size of the parent phase as a consequence of the fact that the transformation proceeds laterally from the grain boundaries of the parent phase.* When the transformation is not complete then the defects are decorated by the presence of the product phase as can often be observed in the case of incomplete oxidation.[33]

6. Radiative effects at defects.

It is well known that grain boundaries act as sources and sinks of point defects. This observation implies that there are segments of edge dislocation lines at grain boundaries that act as such sources and sinks of point defects. For these boundaries to migrate point defects would have to be provided to allow the dislocation segments to climb. Normally, these point defects are generated thermally and diffuse from source to neighboring sink along the grain boundary. However, it is possible to provide point defects via the process of particle bombardment of sufficient energy to produce displacement spikes. Thus, it should come as no surprise that grain growth rates have been increased by ion bombardment.[41] The enhancement in growth rate in this case was found to depend upon the defect yield per incident ion.

Enhancement of grain growth has been found to occur in polycrystalline silicon[42] as a result of light illumination at a level of 4–6 W/cm². This is an interesting result for reasons that will be made apparent below. The investigators claimed that the illumination introduced vacancies that acted to enhance the migration rate of the grain boundaries. This writer believes this explanation unlikely and suggests the following explanation. The absorption of photons in semiconductors is known to produce hole-electron pairs. These charge carriers diffuse through the lattice until they recombine or are trapped. Grain boundaries act as traps and recombination centers for the charge carriers. In the process of non-radiative recombination at grain boundaries the energy released by the recombination produces phonons locally and thereby raising the local temperature. This writer proposes that the enhancement of grain boundary migration rates in polycrystalline silicon resulting from illumination by broad-band light is a consequence of a rise in the local temperature at grain boundaries due to the non-radiative recombination of light generated holes and electrons. Here is one proposed effect where the non-radiative recombination of charge carriers is beneficial rather than detrimental!

The effect of recombination enhancement of transport processes has been noted in the literature. In particular, recombination enhanced dislocation climb (REDC) or recombination enhanced dislocation glide (REDG) have been invoked as a necessary process in the phenomenon of rapid degradation of GaAlAs/GaAs lasers[43] and in II–VI laser structures.[44] The glide of dislocations that are oriented parallel to <110> and become dark line defects (DLDs) is believed to be aided by REDG, whereas the climb of <100> dislocation DLDs is believed aided by REDC which acts to help generate the required duo of Ga and As vacancies and/or interstitials. Since there is a direction to the glide or climb of the dislocation DLDs these processes must be affected by internal stress acting to move the dislocations in a particular direction. Further, the activation energies should be reduced by the internal stress and the process of activation aided by an increase in the local temperature due to recombination near or at the defects limiting the glide or climb of the dislocation DLDs. Recombination enhanced interstitial migration of impurity atoms has been observed.[45] Also, recombination induced dissociation of C–H complexes has been found.[46] Thus, recombination enhanced activated reactions and diffusion are plausible and probable processes in many semiconductors.

There are many cases where it is desirable to reduce the maximum temperature a film will experience in deposition and post-deposition processing. Should the film be a semiconductor then the effect just described may help accomplish this objective. For example, decrease of the minimum epitaxial deposition temperature has been pursued as a goal for many years. Would it not be worthwhile, given the previous discussion and this writer's theory of the factors governing the minimum epitaxial deposition temperature (see *Volume I of Materials Science in Microelectronics*, p. 203) to attempt a reduction of the minimum epitaxial deposition temperature of silicon using light illumination during deposition? This writer can think of many other applications where enhancing the local temperature at line and planar defects can be beneficial. Surely, the reader can also.

Recapitulation

Diffusion along grain boundaries when present affects the time to failure during electromigration of interconnections. For a given area of grain boundary diffusion paths per cross-sectional area of interconnection the rate of mass transport is affected by alloying elements that act to bind vacancies, by the character of the grain boundaries and by the net force acting to drive the net vacancy motion. Not only does the diffusion of vacancies along grain boundaries driven by the electron wind affect the time to failure of the interconnection, but it also acts to increase the resistance of the interconnection through the formation of voids along the latter. Grain boundary diffusion is significant in various processes associated with the production of integrated circuits. For example, it is the primary path for diffusion involved in the formation of an intermediate phase between two reactant layers and, consequently, determines the temperature at which the reaction will proceed to completion in a reasonable time. Also, it is usually the reaction path that produces hillocks on the surfaces of films under compressive stress during processing. Surface diffusion is also often involved in surface roughening and smoothening processes, as well as being necessary for the deposition of various types of films. Structure affects the efficacy of diffusion barriers in various ways. For example, amorphous materials generally tend to have lower diffusivities than crystalline materials having the same melting temperatures. Structure is also significant in solid state reactions in thin films by providing the sites where heterogeneous nucleation occurs preferentially and thereby controlling the grain size of the resulting product phase. Defect structure can also be involved in other processes, such as grain growth, and be affected by other defects produced by irradiation that annihilate at the original defect structure.

References

1. V.T. Borisov, V.M. Golikov and G.V. Scherbedinsky, Phys. Metals Metallogr. 17, 80(1964).
2. D. Gupta, in **Diffusion Phenomena in Thin Films and Microelectronic Materials**, eds. D. Gupta and P.S. Ho, Noyes Publications, 1988.
3. D. Gupta and R. Rosenberg, Thin Solid Films. 25, 171(1975).
4. A.J. Blodgett and E.R. Barbour, IBM J. Res. Devel. 26, 30(1982).
5. I. Kaur and W. Gust, in **Handbook of Grain and Interface Boundary Diffusion Data**, Ziegler Press, Stuttgart, 1989.
6. D. Gupta, MRS Symp. Proc. 47, 11(1985).
7. E.D. Hondros and M.P. Seah, in **Physical Metallurgy**, 3rd edition, eds. R. Cahn and P. Haasen, North-Holland Physics Publications, New York, 1983.
8. I.A. Blech, J. Appl. Phys. 47, 1203(1976); Erratum, J. Appl. Phys. 48, 2638(1977); I.A. Blech and C. Herring, Appl. Phys. Lett. 29, 131(1976).
9. C.A. Ross, MRS Symp. Proc. 225, 35(1991).
10. C.-C. Lee, E.S. Machlin and J. Rathore, J. Appl. Phys.71, 5871(1992).
11. S. Bader, P.A. Flinn, E. Arzt and W.D. Nix, MRS Symp. Proc. 309, 249(1993).
12. P. Gas and F. d'Heurle, MRS Symp. Proc. 402, 39(1996).
13. T.L. Hylton, K.R. Coffey, M.A. Parker, and J.K. Howard, Science. 261, 1021(1993).
14. P. Shewmon, in **Diffusion in Solids**, McGraw-Hill, New York, 1963.
15. H.J. Wollenberger, in **Physical Metallurgy**, 3rd edition, eds. R.W. Cahn and P. Haasen, Elsevier Science Publications, Amsterdam, 1983.
16. See Ref. 11 and S.L. Lehoczky, J. Appl. Phys. 49, 5479(1978).

17. D.R. Srolovitz, W. Yang and M.G. Goldiner, MRS Symp. Proc. 403, 3(1996).
18. C.L. Briant, R.H. Wilson, L. Bigio and W.G. Morris, ibid., p. 21.
19. C.Y. Chang and R.W. Vook, MRS Symp. Proc. 225, 125(1991).
20. J. Falta,T. Schmidt, A. Hille and G. Materlik, Phys. Rev. B54, R17288(1996).
21. S. Takayama and N. Tsutsui, MRS Symp. Proc. 403, 645(1996).
22. C.A. Pico, J. Tao and N. Cheung, MRS Symp. Proc. 225, 113(1991).
23. R. Plugaru, E. Vasile, P. Cosmin, S. Cosmin, C. Cobianu and D. Dascalu, MRS Symp. Proc. 403, 339(1996).
24. S.B. Herner, H.-J. Gossmann and K.S. Jones, MRS Symp. Proc. 402, 143(1996).
25. M.-A. Nicolet, Thin Solid Films. 52, 415(1978); M.-A. Nicolet and M. Bartur, J. Vac. Sci. Technol. 19, 786(1981); C.Y. Ting and M. Wittmer, Thin Solid Films. 96, 327(1982); M. Wittmer, J. Vac. Sci. Technol. A2, 273(1984); S.-Q. Wang, MRS Bulletin XIX (8), 30(1994).
26. C.H. Choi, G.A. Ruggles, A.S. Shah, G.C. Xing, C.M. Osburn and J.D. Hunn, J. Electrochem. Soc. 138, 3062(1991).
27. C.S. Choi, Q. Wang, C.M. Osburn, G.A. Ruggles and A.S. Shah, IEEE Trans. on Elect. Dev. 39, 2341(1992).
28. P. Gas, V. Deline, F.M. d'Heurle, A. Michel and G. Scilla, J. Appl. Phys. 60, 1634(1986); M. Setton and J. Van der Spiegel, J. Appl. Phys. 69, 994(1991).
29. G. Sade and J. Pelleg, MRS Symp. Proc. 402, 131(1996).
30. G. Bai, S. Wittenbrock, V. Ochoa, R. Villasol, C. Chiang, T. Marieb, D. Gardner, C. Mu, D. Fraser and M. Bohr, MRS Symp. Proc. 403, 501(1996).
31. C.K. Huang and S.-Q. Wang, MRS Symp. Proc. 403, 495(1996).
32. F.M. d'Heurle, MRS Symp. Proc. 402, 3(1996).
33. C. Caragianis-Broadbridge, B.L. Walden, J. Blaser, C.O. Yang and D.C. Paine, MRS Symp. Proc. 357, 213(1995).
34. J.A. Kittl, D.A. Prinslow, P.P. Apte and M.F. Pas, MRS Symp. Proc. 402, 269(1996).
35. K. Fujii, R.T. Tung, D.J. Eaglesham, K. Kikuta and T. Kikkawa, ibid., p. 83.
36. R.W. Mann, L.A. Clevenger, G.L. Miles, J.M.E. Harper, C. Cabral Jr, F.M. d'Heurle, T.A. Knotts and D.W. Rakowski, ibid., p. 95.
37. V. Rao and W. Zagozdzon-Wosik, MRS Symp. Proc. 405, 345(1996).
38. S.M. Hu, J. Appl. Phys. 45, 1567(1974).
39. C.-K. Hu, K.P. Rodbell, T.D. Sullivan, K.Y. Lee and D.P. Bouldin, IBM J. Res. Develop. 39, 465(1995).
40. C.A. Cabral Jr, L.A. Clevenger, J.M.E. Harper, F.M. d'Heurle, R.A. Roy, C. Lavoie, K.L. Saenger, G.L. Miles, R.W. Mann and J.S. Nakos, Appl. Phys. Lett. Accepted for publication 1997.
41. H.A. Atwater, C.V. Thompson and H.J. Smith, J. Appl. Phys. 64, 2337(1988).
42. W. Chen, B.L. Sopori, N.M. Ravindra and T.Y. Tan, MRS Symp. Proc. 485, (1998).
43. P.M. Petroff and L.C. Kimerling, J. Appl. Phys. 29, 461(1976); S. O'Hara, P.W. Hutchinson and P.S. Dobson, Appl. Phys. Lett. 30, 368(1977).
44. S. Tomiya, M. Uketa, H. Okuyama, K. Nakano, S. Itoh, A. Ishibashi, E. Morita and M. Ikeda, Mat. Sci. Forum. 196–201, 1109(1995).
45. H. Nakashima, T. Sadoh and T. Tsurushima, ibid., p. 1351.
46. Y. Kamiura, F. Hashimoto and M. Yoneta, MRS Symp. Proc. 262, 549(1992).

Bibliography

Grain Boundaries: A.P. Sutton and R.W. Baluffi, **Interfaces in Crystalline Materials**, Oxford University Press, 1995.

Diffusion in Semiconductors: Concise Encyclopedia of Semiconducting Materials & Related Technologies, eds. S. Mahajan and L.C. Kimerling, Pergamon Press, 1992, see articles starting on pp. 102 and 108.

Diffusion in Solids: R.J. Borg and G.J. Dienes, **An Introduction to Solid State Diffusion**, Academic Press, 1988.

Thermodynamics: E.S. Machlin, **An Introduction to Aspects of Thermodynamics and Kinetics Relevant to Materials Science**, GiRo Press, 1991.

Appendix

A1. Electromigration in copper conducting lines.

In the previous decade (<2005) aluminum copper conducting lines have been replaced with copper. The rationale for this replacement stems from the miniaturization trend, the need to achieve higher conductivity and the corresponding need to enhance the electromigration resistance in view of the increase in current density. Copper has higher conductivity than aluminum and the grain boundary diffusivity of copper is orders of magnitude smaller than that for aluminum. Hence, the choice of copper to replace aluminum seemed eminently reasonable. However, the fact is that problems have arisen that were not foreseen that are responsible for electromigration induced failures of copper conducting lines at lifetimes much smaller than expected.

One such problem arose because the Cu diffusion path induced by the current turned out to be the interfaces of the conducting line with its surroundings instead of the grain

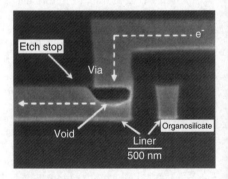

Figure A5.1. A void underneath a via. A layer acting as a barrier to mass transport separates the bottom of the via from the bottom Cu line. Reprinted with permission from J. He et al., Appl. Phys. Lett. 85, 4639(2004). Copyright 2004 American Institute of Physics.

boundaries. In particular, the interface between the Cu line and its diffusion barrier was the active diffusion interface. It was found that providing for a metallurgical bond between the diffusion barrier and Cu greatly reduced diffusion along this interface.[A1] (The interface diffusivity in the presence of a metallurgical bond approaches that of a grain boundary rather than that of a surface.)

Figure A5.1 shows the result of a failure in design in a dual damascene copper line. A layer acting as a barrier to the mass transport of copper is present at the bottom of a via connecting two lines. Obviously, a divergence in the mass transport of Cu occurs at this layer bringing about the production of a void-an open in the line-a failure. Removal of this barrier solved this problem.

Because of the need to minimize dimensions the shunt layer normally present in the Al–Cu technology is not used in the Cu damascene technology. The consequence of this lacuna is that only one void along a line will induce failure of the line. With a shunt line, the shunting of the void allowed the line to continue conducting current and avoided an early failure. Hence, it is important to design the Cu lines so as to eliminate as much as possible any divergences in the mass transport induced by the current. Another possible solution is to design the line lengths to be smaller than the length at which the back stress cuts off mass flow.

Reference to Appendix

A1. C.-K. Hu et al., Appl. Phys. Lett. 81, 1782(2002).

Problems

1. What characteristics in terms of structure define a good diffusion barrier material?
2. In an interconnect line containing a bamboo structure not all the grain boundaries are normal to the interconnect line but there are some that connect two bamboo type grain boundaries. Describe what will happen and where it will happen to this interconnect line during electromigration.
3. Which type of grain boundary would have a lower self-diffusivity – one located at a cusp in the energy or one located away from a cusp?
4. How would you attempt to reduce the propensity to form hillocks on a thin film?
5. Why is it that the grain size of a product phase produced by the reaction between two layers often has the same grain size as the parent phase it replaces?
6. Outline the steps you would take in calculating the local temperature increase at a grain boundary, in a one micron thick polycrystalline silicon film deposited upon a glass substrate, due to irradiation by broad-band light.

CHAPTER VI

Interface and Junction Properties

1. The effect of structure on junction properties.

1.1. Schottky junctions.

The ability to predict the Schottky barrier height (SBH) for all metal–semiconductor junctions has eluded scientists to date. One reason for this state of affairs is that for many junctions the structure and composition of the interface region is not known. To understand why this knowledge is necessary it is helpful to consider Schottky's original model for this junction as follows. Consider the case of a metal which has a work function, Φ_M, larger than the electron affinity, χ_S, of an n-type semiconductor, where these quantities are defined as the difference between the vacuum level and the Fermi levels of the electron in the bulk materials, respectively. Let us assume that relative to the vacuum level the conduction electrons in the metal have a lower chemical potential than those in the semiconductor. Thus, when contact is made between the two materials electrons will transfer to the metal building up at the interface until the chemical potential is equal in the two components of the junction. This charge derives from the near-surface region of the semiconductor adjacent to the metal. Figure 6.1 taken from Phillips's book[1] shows an electron energy-distance diagram for various steps in the process of making the junction under the assumption that a wire of decreasing length connects the metal and semiconductor. As shown there the consequence of the contact is an energy barrier between metal and semiconductor with height equal to

$$\Phi_{Bn} = \Phi_M - \chi_S \qquad (6.1)$$

However, in fact, the SBH of metal–semiconductor junctions does not obey equation (6.1). The reason for this aberrant behavior is that the *work to remove an electron from the surface of either the real metal or semiconductor does not equal necessarily that to remove an electron from the respective bulk Fermi levels.* The sensitivity of the work function of materials to their surface composition is well known, e.g. the work function of a tungsten surface having a Cs monolayer on it is much smaller than that from a clean tungsten surface, etc. Thus, this concept should not be surprising. Indeed, band bending at semiconductor surfaces due to surface charge at surface states or surface dipole layers does occur and affects the work function. Further, even if the surfaces of both metal and semiconductor were clean and the surface structure was such that no band bending at the surfaces existed prior to contact between them there is no reason to expect that the atomic configurations of the surface planes would remain unchanged upon contact. Thus, the formation of a metal–semiconductor interface should be expected to affect the band bending, at least on the semiconductor side of the interface.

How it affects the band bending is the subject matter of the many different theories of the Schottky barrier height. One of the latest concepts advanced to answer the question just raised is that after equilibration of the bulk Fermi levels, and in the absence of all other effects on band bending at the surface, the energy level corresponding to the charge neutrality level of the semiconductor, at its interface with the metal, should be at the common Fermi level. Now there are at least two views of the factors that govern the charge neutrality level. According to Tersoff[2] the charge neutrality level corresponds to the branch point in the virtual gap states of the semiconductor, that are induced by the tails of the metal wave functions that penetrate into the semiconductor. These states are now known as metal-induced-gap states (MIGS).[3] At this branch point the MIGS change from donor to acceptor character. According to Lanoo and Friedel[4] the charge neutrality level occurs at the average of the dangling bond levels for the atoms at the idealized unrelaxed, unreconstructed surface of the semiconductor that is joined to the metal. These dangling bond states resonate with the MIGS. These concepts are applicable to ideal surfaces between

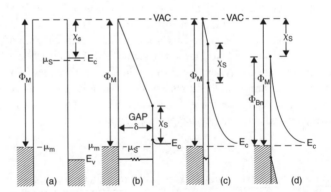

Figure 6.1. Band bending at a Schottky metal-n-type semiconductor junction. In (a) the metal and semiconductor are separated, and all energies are measured relative to vacuum. In (b) the samples are joined by a wire, which makes the chemical potential of the electrons equal in both components. There is a potential drop in the gap of $\Phi_M - \chi_S$. As the gap is reduced to a few microns substantial surface change accumulates on the metal and semiconductor, and some of this spreads out into the interior of the semiconductor, causing the band-bending shown in (c). Finally, upon ideal contact, band-bending accounts for all the barrier height as shown in (d). Reprinted with permission from J.C. Phillips, **Bonds and Bands in Semiconductors**, p. 254, Copyright 1973 by Academic Press, Inc.

which there is no charge transfer other than that to equilibrate the bulk Fermi levels. It is somewhat surprising that all the different ways of calculating the position of the charge neutrality level yield nearly the same value for the different semiconductors.[5]

Now surfaces are seldom ideal and charge transfer can occur between metal and semiconductor. Monch[6] accounts for the charge transfer due to an electronegativity difference with the following relation for the case of an n-type semiconductor

$$\Phi_{Bn} = \Phi_{cnl} + S_X(X_m - X_s) \quad (6.2)$$

where Φ_{cnl} equals the difference, at the metal–semiconductor interface, between the energy of the bottom of the semiconductor conducting band and the charge neutrality level, X_i is the appropriate electronegativity of the i component of the junction and S_X is a proportionality parameter that is solely a function of the properties of the semiconductor. We will return to consider this relationship shortly, but before we do let us consider how Monch evaluates X. He makes use of the Miedema electronegativities,[7] but applies them as if the values calculated for the bulk (for a compound $A_M B_N$ equal to $(X_M^m \cdot X_N^n)^{1/(n+m)}$) hold for the surface. But experimental evidence demonstrates that the charge transfer is local between contacting atoms only.[8] We will provide an explanation later why use of these bulk averaged electronegativity values appear to yield behavior consistent with equation (6.2), such as is shown in Figure 6.2. The upper line corresponds to equation (6.2) in that when the electronegativity difference is zero the SBH should equal Φ_{cnl} and, in fact, a point corresponding to a theoretical value of the latter parameter, as calculated by Tersoff,[2] lies near this line. (Values of Φ_{cnl} have been tabulated by Monch.[9]) The spread of experimental points below this line indicates that the SBH is sensitive to additional influences not included in equation (6.2).

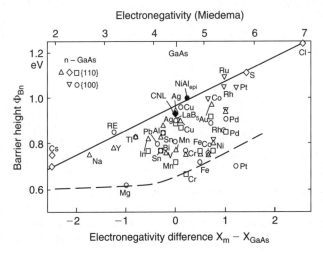

Figure 6.2. Barrier heights for metal–GaAs contacts versus Miedema electronegativity difference. Reprinted with permission from W. Mönch, Surf. Sci. 300, 928. Copyright 1994 by Elsevier.

Although Monch considers the effects not included in equation (6.2) to be secondary, others consider them to be primary, in accounting for the deviation from ideal Schottky behavior.[10] Among the latter are the effects of interface defects, interface dopants, interface structure, interface compound formation, interdiffusion and of bonding across the interface. Indeed, using as a measure the deviation from the ideal SBH, we will later show effects on SBH both larger and smaller than that due to pinning of the Fermi level at the charge neutrality level of the MIGS. However, Monch does not deny that these effects can exist, nor that, in principle, they need be quantitatively smaller in their effect on SBH.

All of these additional effects on SBH correspond to effects of structure and, hence, are those that are of primary interest to us. Some of these effects act to add or detract charge which must be considered in evaluating the charge neutrality condition. Monch's analysis of these additional effects is as follows. The charge neutrality condition is given by

$$Q_m + Q_{id} + Q_{migs} + Q_{sc} = 0 \qquad (6.3)$$

where Q_m is the charge density residing in the metal, Q_{id} is the charge density residing in the interface defects on the semiconductor side, Q_{migs} is the charge density residing in the MIGS in the semiconductor, and Q_{sc} is the space charge density in the depletion layer in the semiconductor, which is ordinarily negligible compared to Q_{migs} and will be neglected in the following. For a constant density of states, D_{migs}, in the MIGS

$$Q_{migs} = eD_{migs}(\Phi_{Bn} - \Phi_{cnl}) \qquad (6.4)$$

The latter quantity in parentheses equals the difference between the energy of the charge neutral level and the Fermi level. Now according to Monch the total charge density transferred to the metal, $-Q_m$, equals $S_X(X_m - X_S)$, i.e. is a constant for a given metal–semiconductor junction. Hence, for $X_m > X_S$, Q_m is a negative charge density. Hence, by equations (6.2–6.4) there must be a change in Φ_{Bn} equal and opposite to Q_{id}/eD_{migs}, i.e.

$$\delta\Phi_{Bn} = -Q_{id}/eD_{migs} \qquad (6.5)$$

Consequently, on an n-type semiconductor positively charged interface defects decrease the SBH and negatively charged interface defects increase the SBH. It is expected that the maximum interface defect charge density is some fraction of the product of an electronic charge by the total number of atoms per unit area of interface. Monch has interpreted the lower envelope in a plot such as that of Figure 6.2 to represent the limit to $\delta\Phi_{Bn}$ due to the maximum interface defect charge density.

Interface dopants and interface structure introduce dipoles that affect the SBH. An example of the former effect is that due to a Schottky junction formed between a metal and a hydrogen-terminated p-diamond surface as compared to one formed on a clean p-diamond surface with the results given in Figure 6.3. (The barrier

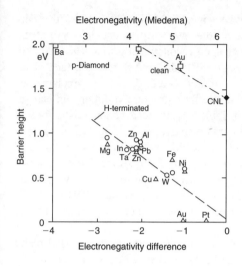

Figure 6.3. Schottky barrier height vs Miedema electronegativity difference for clean and hydrogenated p-diamond–metal junctions. Reprinted with permission from W. Mönch, Europhys. Letts. 27, 479. Copyright 1994 by EDP Sciences.

height in this figure is Φ_{Bp}.) In this case, the most electronegative component is the semiconductor, diamond, and the adsorbed hydrogen layer produces a dipole of the same polarity as that produced by the electronegativity-induced transfer of charge from the metal to the semiconductor thereby decreasing Φ_{Bp}. The same conclusion can be reached by considering the dopants at the interface to be interface defects with the charge on the dipole plane nearest to the semiconductor side of the interface. Thus, for the junction between metals and hydrogen-terminated diamond, the sign of the dipole charge in the semiconductor is negative and by equation (6.5) the value of $\delta\Phi_{Bn}$ would increase. Hence, $\delta\Phi_{Bp}$ should decrease as it does. Similarly, on n-GaAs, electropositive Ti, Mn and Al adsorbants should produce a positive charge facing the semiconductor side, at the interface between the semiconductor and Ag, Au and Pd contacts (the latter are more electronegative than the adsorbants or semiconductor), and thereby act to reduce the SBH while electronegative S, Se and Te adatoms relative to the metal contacts present a negative charge on the semiconductor side of the dipole and thus act to increase the SBH, as is observed.[11]

At this point it is appropriate to explain why the Miedema average electronegativity appears to work for compound semiconductors, as shown in Figure 6.2, despite the fact that many of the surface planes, on which metal was deposited to produce the Schottky junction, *consist ideally only of atoms of a single element*. All the data were obtained for junctions made by deposition at room temperature. The mobility of atoms at surfaces at room temperature is sufficient to allow for intermixing to occur during deposition when there is only a fraction of a monolayer of metal on the semiconductor surface so that metal atoms contact both components of the semiconductor across the now diffuse interface and, consequently, use of the average electronegativities is appropriate.

Interface structure dipoles have been considered by Monch[12] with a correct prediction of the shift in SBH due to the removal of a stacking fault at the 7×7 Si(111) interface with a Ag layer upon reconstruction to a 1×1 Si(111) interface. On the other hand, the changes in SBH due to different interface structures in the NiSi$_2$/Si junction system[13] were not predicted, but were explained in terms of assumed dipole polarities. That interface structure variation can change the SBH is indicated by the data listed in Table 6.1. Just what characteristics of this structure variation control the SBH are not known. However, Tung[10] has suggested that dipoles brought about by changes in metal and semiconductor bonding hybridizations at the interface account for the major effect of this interface structure variation. Band structure calculations[79] based on the local density approximation using large supercells verify this suggestion and produce SBH results in agreement with the SBH values measured on the A and B interfaces of the NiSi$_2$/Si(111) junctions listed in Table 6.1.

Further studies of the effect of the semiconductor surface crystal plane at the metal–semiconductor interface reveal large changes in SBH from 0.2 to 0.85 eV with change in the orientation of this plane away from (100) for an Al/GaAs junction. In this case the SBH variation is monotonic with change in the number density of steps along the (100) plane as the misorientation increases.[80] However, this variation may be a consequence of a dipole due to metal–cation replacement along these ledges on the compound semiconductor surface. A similar effect has been observed on a Sc$_{0.32}$Er$_{0.68}$As/GaAs junction as the surface orientation of the GaAs plane is changed from (100) to (111) rotating about $[01\bar{1}]$ with a concomitant variation of SBH from 1.03 to 0.63 eV.[81] These changes are noteworthy because they span about half the band gap.

Table 6.1. Interface structure dependence of Schottky barrier height

Junction	On n-Si, eV	On p-Si, eV
NiSi$_2$/Si(001)[13+]	0.4	
NiSi$_2$/Si(111)A[13+]	0.65	
NiSi$_2$/Si(111)B[13+]	0.79	
CoSi$_2$/Si(111)B-1 $\times$ 1[14+]	0.69	0.45
CoSi$_2$/Si(111)B-$\sqrt{3} \times \sqrt{3}$[14+]	0.27	0.71+
CoSi$_2$/Si(001)-Co(6)[15]	0.78	
CoSi$_2$/Si(001)-Co(4/8)[15]	0.67	
PtSi/Si(001)[16,17]	0.88	0.216
PtSi/Si(111)[18]		0.292
IrSi$_3$($2\bar{1}10$)//Si(111)[19+]	0.98	0.12

+ Only these junctions are uniform along their area. The remainder not denoted with the + symbol are non-uniform junctions consisting either of a mixture of textures or a mixture of interface types.

Effects due to interface defects per se are ambiguous in interpretation. Indeed, it is difficult to isolate such effects because the production of defects invariably involve other changes at the interface. One such effect on SBH is due to a variation of misfit dislocations arranged parallel to the metal–semiconductor interface. Each misfit dislocation contains charge which gives rise to a depletion region extending radially outward from the dislocation line. As misfit density increases, depletion regions overlap and increase the barrier height over a greater portion of the interface area.[82]

Monch has also investigated the slope factor S_X and he has found that it depends primarily upon the electronic susceptibility of the non-metal component of the junction. For ionic crystals, S_X takes on the value that yields equation (6.1) as the best description of the SBH, i.e. S_X approaches, A, the constant that relates the work function (and electron affinity) to the electronegativity in the relation

$$\Phi_M = B + AX_M \qquad (6.6)$$

or to the electron affinity by $\chi_S = C + AX_{SC}$. This slope parameter becomes smaller than A for semiconductors, such as Si.

In summary, the structural dependence of the SBH is manifested via the contributions that yield deviations from equation (6.2), among which are interface defects, interface dopants (adsorbants), interface structure, interface compound formation and interface intermixing.

1.2. Ohmic junctions.

The property of ohmic junctions that may be sensitive to structure is the contact resistance. Very often ohmic contacts are made using materials that yield a barrier to the passage of charge carriers, i.e. a Schottky junction. In order to counteract the effect of this barrier dopants are used to encourage tunneling through it. Structure enters the equation in several ways. One is via the dopant distribution, which must be high at the contact. For example, for $CoSi_2$ or $TiSi_2$ ohmic contacts to Si the contact resistance is sensitive to the boron concentration at the silicide–silicon interface. Any process that affects this concentration affects the contact resistance. In particular, it has been found that pre-amorphization to induce the C49 to C54 structure of $TiSi_2$ decreases the boron concentration at the interface with a consequent increase in the contact resistance.[20] Another way structure can affect the contact resistance is via the morphology at the interface. Sometimes, the roughening of the interface under these circumstances can lower the specific contact resistance by increasing the contact area. Such roughening can be induced by interaction between the metal contact and the semiconductor. However, such interaction is usually undesirable because it often brings about deleterious effects on the behavior of the device associated with the junction. Diffusion barriers are then used to prevent or hinder this interdiffusion. Another method of lowering the contact resistance when barriers control the current is to use contact materials that lower the barrier height. For example, the use of an intermediate layer of small band gap, such as $In_xGa_{1-x}As$, between the metallization and p-InP replaces a single barrier with two smaller barriers. However, even in this case it is necessary to dope both semiconductors at the interface to minimize the effect of the presence of such barriers.[21] Dopants, which are integral components of the metallic contact, offer the advantage that their presence via interdiffusion across the interface, a phenomenon that is difficult, if not impossible, to prevent in the striving of the system to reach thermodynamic equilibrium, is beneficial rather than being detrimental to the contact resistance.[22]

There are contacts where the barrier heights are near zero as occur in some cases when the bond strength between the anion component of the semiconductor and the metal contact is stronger than that between the cation and anion components of the semiconductor. The contact resistance in this case seems to depend either upon the state of the interface (i.e. the presence or absence of foreign species, intermediate phases, such as oxides and, perhaps, the structure at the interface), or the state of either contact at the interface. For example, it has been suggested that the stoichiometry at the interface of the compound conductor InP markedly affects the contact resistance to a Ni contact[23] with the resistance decreasing as the stoichiometry approaches unity. It is well known that the presence of defects can increase the resistance by various mechanisms. Defects can act as recombination centers, charge compensation centers, etc. Thus, their absence usually minimizes the contact resistance and care is normally taken in processing to reduce the defect concentration as much as possible. In both types of ohmic contact the contact resistance seems to be lowest when there is no intermediate phase at the interface, when the interaction between the components of the junction (other than that due to doping) is minimal and when the interface is sharp and bonding across it is complete along the entire interface. This observation is the basis for the search for contacts that are in thermodynamic equilibrium.

2. Interfaces and properties.

2.1. Semiconductor–insulator interfaces.

2.1.1. Si/SiO$_2$.

Interface structure and roughness directly affect the interface state density and the properties of devices such as MOS gates in silicon technology. The interface structure refers to the defect structure at the interface between silicon and silicon dioxide. The effect of this interface on breakdown voltage has been discussed in Chapter I and two of the defects appearing at the interface have been described in Figure 1.15. An acceptable interface trap density is less than about $10^{11}/cm^2/eV$, which implies that in this case at most one in every 10^4 to 10^5 Si atoms is not bonded to the oxide or does not have a dangling bond that is not terminated by hydrogen or other passivating species (i.e. the interface defect density as determined electrically is not sensitive to the number of hydrogen terminated dangling bonds but is sensitive to the number of unterminated (unpassivated) dangling bonds). Terminated and unterminated dangling bonds are not the only defects present at the Si/SiO_2 interface. Among the other types of defects at the interface, are strained bonds[24–26] and defects at the interface associated with the coordination of the atoms there (i.e. the local composition).[27,28] We shall examine these and other interface defects in more detail later.

There are two distinct direct effects of the interface states on properties: one on the surface potential and another on the effective mobility in the channel region. These parameters affect the transconductance, the threshold voltage and other properties of field effect transistors. Some of the latter properties are also affected by the quality of the oxide itself. Measureable detrimental effects of the interface defect density on these parameters occur above about $10^{11}/cm^2/eV$.

The main indirect effect of interface defects is on the degradation of other properties, such as breakdown voltage and leakage current with use of the device as already considered in Chapter I. For MOS devices, DiMaria[29] and others[68] have shown that the dominant defect generation mechanisms are associated with H-atom *motion* to the Si/SiO_2 interface where the H atoms recombine with the terminating H atoms to produce H_2 and unterminated dangling bonds. Thus, processing procedures that contribute potentially mobile H atoms to the SiO_2 insulator are expected to yield high interface defect densities under stress conditions.

Apparently for the mobile H to be able to produce a net increase in the number of dangling bonds at H-terminated bonds rather than to terminate dangling bonds at the Si/SiO_2 interface there must be many more of the former defects present at the interface than of the latter ones. We may from the work of Bhat and Saraswat[30] obtain estimates of limiting and minimum hydrogen concentrations in the various oxides they examined. From the fact that the change in the mid gap interface state density for a thermally grown dry oxide due to electric field stressing (300°C for 10 min up to fields of ± 6 MV/cm) is less than the sensitivity of $10^{10}/eV/cm^2$ we may conclude that this oxide had less than about 10^{10} mobile H atoms in the oxide during the stress period in a volume given by the product of $1\,cm^2$ by the oxide thickness. The thickness of this oxide was 21.2 nm. Hence, a density of less than about $5 \cdot 10^{15}/cm^3$ of mobile H atoms was produced during the stressing period from the volume of the oxide or less than $10^{10}/cm^2$ from either interface. If we assume that the mobile H are released from their traps by energetic charges then these numbers represent the limiting densities of trapped hydrogen in these sites for this oxide. The total mid gap interface state density for this thermally grown dry oxide was $3.5 \cdot 10^{11}/eV/cm^2$. Hence, the thermally grown dry oxide contained fewer H-terminated dangling bonds than unterminated dangling bonds at the interfaces with the oxide. However, the conclusion of this analysis differs from that of another study[59] which found that roughly half of the dangling bonds ($5.5 \cdot 10^{12}/cm^2$) at a thermally grown oxide/Si(111) interface were passivated by hydrogen. It must be noted, however, that the former studies were conducted on a Si(100) surface.

For an oxide produced by LPCVD from tetraethylorthosilicate on the other hand, the increase in mid-gap density of states for the same stressing was greater than $10^{12}/eV/cm^2$ and at $-6\,MV/cm$ approached a value of $10^{13}/eV/cm^2$.[30] Hence, the density of trapped H exceeded $5 \cdot 10^{18}/cm^3$ in the latter oxide or $10^{13}/cm^2$ at the oxide/electrode interface. Also, since the measured mid gap interface state density prior to electrical stressing was $2.7 \cdot 10^{11}/eV/cm^2$ then we may conclude that for this oxide the density of H-terminated dangling bonds at the oxide interfaces exceeded by far the density of the unterminated dangling bonds.

Since these oxides were produced on a (100)Si surface the results indicate that it is possible, by a technique such as that of growing a dry oxide thermally, to obtain an interface having no more than $3.5 \cdot 10^{11}/cm^2$ of interface states (i.e. of dangling bonds) at a (100)Si/SiO$_2$ interface. This result suggests that along a planar section of such an interface it is possible to achieve complete bonding without the presence of dangling bonds. The dangling bonds at the interface are probably localized at steps on the Si side of the interface. This deduction is supported by second-harmonic generation studies which reveal different symmetry contributions from dangling bonds from different regions on the surface[47] and by studies of the effect of step atom density from LEED spot broadening which showed that the density of interface traps was proportional to the square of this step density. (This conclusion is at odds with a conclusion of Stesmans and van Gorp[59] that the dangling bonds are not located at steps of a Si(111) surface but are on the terraces). Apparently, a high temperature annealing step (>900°C) can act to smoothen a rough surface and also to enable SiO$_3$ groups to bridge dangling bonds at these steps for temperatures above 1000°C.[47] However, too high an annealing temperature, which allows oxygen to diffuse into the silicon from the Si/SiO$_2$ interface, under the thermodynamic driving force to produce the oxygen concentration in the Si that is in equilibrium with SiO$_2$, also can produce dangling bonds along the planar section of the interface. Thus, the Si/SiO$_2$ interface is blessed with the possibility of continuity of bonding across an ideal planar interface and, in principle, an absence of dangling bonds, as illustrated, for example, in Figure 6.4 for the (111)Si surface.

Summarizing, the main type of defect present at the interface between Si and thermally grown SiO$_2$, and which is measured electrically in evaluation of the

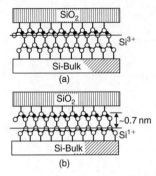

Figure 6.4. Atomically flat Si(111)/SiO$_2$ interface structures for two extreme cases: (a) interface structure consists only of Si^{3+}; (b) interface structure consists only of Si^{1+}. Reprinted with permission from T. Hattori and K. Ohishi, MRS Symp. Proc. 318, 61(1994).

interface state density, is the dangling bond, P$_b$, defect.[67] Two important mechanisms of interface degradation involve the P$_b$ defect and are hot-electron-related and trapped-hole-related. Hot electrons with energies greater than 2 eV above the oxide conduction band edge cause the release of hydrogen at the anodic interface. This hydrogen then diffuses to the cathodic interface and depassivates already passivated P$_b$ defects to produce H$_2$ and the dangling bond, the P$_b$ defect. If electron injection is from the silicon substrate and exceeds a critical fluence, as is common in MOSFET devices, this process takes place at the Si/SiO$_2$ interface causing threshold voltage shifts and mobility degradation. In the case of trapped holes, the latter act to crack hydrogen and generate H$^+$, which will drift to the Si/SiO$_2$ interface under positive bias conditions and depassivate already passivated P$_b$ defects after capturing an electron there.[68]

One of the other type of defects produced by the presence of H has a positive charge and has been denoted as an anomalous positive charge, APC, defect. A mechanism has been proposed[31] for this defect which involves the following reactions at the Si/SiO$_2$ interface:

$$\text{Si—O—Si} + \text{H}^0 + \text{h} \rightarrow (\text{Si—OH—Si})^+$$

$$\text{Si}_2\text{—NH} + \text{H}^0 + \text{h} \rightarrow (\text{Si}_2\text{—NH}_2)^+$$

The trapping of electrons by these APC defects regenerate the starting configurations and release H atoms that can again react with holes and defects to produce these positively charged defects. Thus, the sequence of reactions correspond to a mechanism for interface recombination.

The composition at the Si/SiO$_2$ interface over about one to two monolayers is close to SiO. This fact may be related to the continuity of bonding across planar sections of the Si/SiO$_2$ interface. Certainly, the last word has not been said as yet concerning the atomic arrangements at this interface for the variety of Si orientations that are of interest.

Much has been said in this subsection concerning the structure of the Si/SiO$_2$ interface, but little has been said about the effect of this structure, as measured

Table 6.2.

Split	Oxide type	Anneal conditions		
		1000°C A	600°C O_2	600°C A
DH	Dry O_2 1000°C	20 min	–	–
WN	H_2O 900°C	–	–	–
LH	Silane 400°C	20 min	–	–
LL	Silane 400°C	–	12 h	5 h
LN	Silane 400°C	–	–	–
TH	TEOS 650°C	20 min	–	–
TN	TEOS 650°C	–	–	–

In first letter of split symbol D denotes dry thermal oxide, W denotes
wet thermal oxide, L denotes silane LPCVD grown oxide, T denotes
tetraethylorthosilicate grown oxide; in second letter of split symbol H
denotes 20 min annealing at 1000°C, N denotes states after deposition of
gate electrode at 580°C and annealing at 600°C – a step included for all
samples, and L denotes after annealing for time shown at 600°C.

by interface state density on properties. Tables 6.2 and 6.3 present data from the
experiments of Bhat and Saraswat[30] on the effect of various LPCVD oxide deposition techniques and annealing conditions on the mid gap interface state density
before and after stressing at -6 MV/cm, on the change in flatband voltage due to
such stressing and on the charge to breakdown compared to controls of thermally
grown dry and wet oxides. These data reveal no correlation between the charge to
breakdown or the decrement in flatband voltage to the initial interface state density
prior to stressing. However, the charge to breakdown correlates to the interface state
density after stressing with the exception of that for one sample, the one denoted
LH. For the interface density after stressing of 10^{12}/cm^2 one would expect that the
charge to breakdown would be less than that given in the table. The data indicate
that above an interface state density after stressing of $1.8 \cdot 10^{12}$/cm^2 there are 3 data
points with charge to breakdown $\leqslant 1$ Coloumb/cm^2, as expected. Hence, the writer
believes there may be an error associated with the numbers associated with the
sample denoted LH. Further, the decrement in flatband voltage also correlates to
the interface state density after stressing with some small scatter. The model based
on the reaction between H atoms with both unterminated and terminated dangling

Table 6.3.

Oxide	Interface state density $10^{11}/cm^2$	Increment in state density $10^{11}/cm^2$	Charge to breakdown C/cm^2	Decrement in flatband voltage volts
DH	3.5	<0.1	42.2	<0.01
WN	2.8	3.2	17.1	0.095
LH	3.6	6.8	35.0	0.108
LL	16.0	>2.2	1.0	>0.08
LN	4.4	35	0.6	0.3
TH	2.7	3.2	27.0	0.01
TN	2.7	90	0.02	1.1

bonds at the interface is consistent with the above described data with the exception of the single sample LH. Nevertheless, it would be useful to have independent measurements of the terminating H content at the Si/SiO$_2$ interface.

2.1.2. Compound semiconductor–insulator interfaces.

The fortunate situation of continuity of bonding across the semiconductor–insulator interface described above for Si/SiO$_2$ does not exist for most of the compound semiconductors. Thus, while it is relatively easy to produce a low interface defect density for the Si/SiO$_2$ interface, it is difficult to achieve a similarly low interface defect density for compound semiconductor–oxide interfaces. Thus, the properties of MOSFET devices in compound semiconductors have been inadequate for commercial application. For this reason, MOS structures in compound semiconductors have been replaced by MES structures which make use of the Schottky barrier at the metal–semiconductor interface as a gate in place of the oxide. However, although difficult, a recent result has shown that this objective of a low defect density at a compound semiconductor–oxide interface is not impossible. In particular, Passlack and Hong[32] have produced, by means of MBE deposition of amorphous Ga$_2$O$_3$ onto a surface of GaAs grown by solid source MBE *which had less than 10 Langmuirs exposure to oxygen* prior to oxide deposition, Ga$_2$O$_3$/GaAs interfaces having interface state densities in the low 10^{10}/cm^2/eV range with interface recombination velocities of about 4000 cm/s. Oxide films of Al$_2$O$_3$, SiO$_2$, and MgO similarly deposited, however, did not yield these low interface state densities, which suggests that the bonding at the interface involves a continuation of bonding across the interface for the Ga$_2$O$_3$/GaAs case, and not for the other oxides. Also, the production of the Ga$_2$O$_3$/GaAs interface must have been accomplished without the formation of free As clusters at the interface and the consequent Fermi level pinning at the interface.[33] Thus, perhaps with the further development of this technique of using molecular beams of the cation oxides depositing onto ultra-clean compound semiconductor surfaces it may be possible to produce commercial MOS devices in compound semiconductors.

2.1.3. Passivation of compound surfaces and properties.

Compound semiconductors are often used as substrates for quantum wells and other devices. In many of these applications passivation of the surface states is necessary to reduce their density and effect on non-radiative surface recombination or on leakage and low-field breakdown in electronic devices. In the previous sub-section we noted that most insulators cannot be deposited on compound surfaces without the presence of an undesirable number of detrimental surface states. We also described there one method of depositing an oxide on GaAs without the production of such states. There exists another potential procedure for reducing the density of these defect states which may allow the production of useful metal–insulator–semiconductor devices in the future. This procedure involves the passivation of these detrimental states.

One technique that has been used to passivate the undesired surface states of GaAs is to react the surface with sulfur. Although photoluminescence measurements reveal that such passivation decreases the density of non-radiative recombination centers located at the GaAs surface, this beneficial effect decreases with time.[42] Although many attempts to produce a sulfidized surface that does not degrade with time have been made none have yet been commercially acceptable. Another possible mode of passivation is to make use of hydrogen. In particular, it has been found that hydrogen ion passivation yields stable (at least 3 years) near-surface quantum wells with improved luminescence.[43] The passivation was found to be sensitive to the hydrogen ion dose as revealed in Figure 6.5. Here it is evident that too small a dose is insufficient to passivate all the defects and too large a dose is detrimental. What does passivation in this application correspond to physically? From numerous studies Shi et al.[45] believe that the surface defects responsible for non-radiative recombination are As_{Ga} anti-site defects. When sufficient hydrogen is reacted to remove these anti-site sited As atoms as arsine, without further reaction, then the surface is believed to be passivated. Given the result of Passlack and Hong[32] reported in the previous subsection, it seems to this writer that their technique of limiting the contact of the surface to less than 10 Langmuirs of oxygen will also passivate the surface. Further, in view of the detrimental results of oxygen contact with the surface it appears to this writer that the hydrogen ion beam used by Shi et al. also must have produced a hydrogen terminated surface which protected it from reaction with oxygen.

When another layer is deposited on the GaAs surface the removal of As_{Ga} anti-site defects does not necessarily produce a passivation of the surface *for all types of layers subsequently deposited on this surface*. The removal of these defects assures that there are no dangling bonds at the interface *with a subsequently*

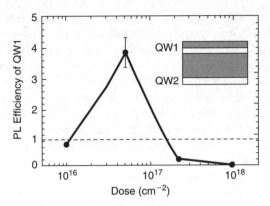

Figure 6.5. Normalized photoluminescence efficiency versus hydrogen ion dose. The PL intensity for upper 70 Å QW1 (see the inset) is normalized to that of deep, 100 Å QW2. That value is subsequently divided by the normalized intensity for a sample not subject to hydrogenation. Reprinted with permission from S.S. Shi, Y.-I. Chang, E.L. Hu and J.J. Brown, MRS Symp. Proc. 421, 401(1996).

deposited layer that continues the bonding pattern across the interface. However, the absence of these defects does not assure the absence of dangling bonds at the interface with an incompatible bonding configuration structure, such as that of Al_2O_3, for example. In the latter case, passivation requires the satisfaction of all dangling bonds at the interface and the production of a surface that will not react with atoms that come into contact with it.

For the case of InP it appears that the native oxide that appears on this surface is actually a mixture of oxides. It has been found that if a thin layer of Si is deposited on a supposedly clean InP surface followed by deposition of a SiO_2 layer that a MIS using the Si/SiO_2 as the insulator can be produced that yields good properties.[69] It has been speculated that the Si layer reduces the native oxides on the supposedly clean InP to form SiO_2 upon annealing and an interface with the InP that is consistent with good properties from the device. It would be interesting to know more about the interface state density in this case and whether the same scheme can be used with other compound semiconductors. The good device results imply that the interface state density is well below $10^{12}/cm^2$ and that there is a good match in bond continuity between SiO_2 and InP. Interface state density measurements for an Al_2O_3/InP interface[70] yield values between 2–$3 \cdot 10^{12}/cm^2$ and it is known that MIS devices from this combination have unsatisfactory properties.

2.1.4. Interfaces and luminescent properties.

The effect of non-radiative recombination centers on various properties has already been mentioned in prior chapters. In the case of porous silicon the passivation of the surface either by hydrogen or oxygen seems to play two roles: removal of non-radiative recombination centers and the production of luminescent centers at the surface. Also, in the cases of Er nanoparticles[34] embedded in a thin film poly-Si matrix rich in oxygen, of Ge[35–37], Si[37–39] and SiC[40] particles embedded in a thin film SiO_2 host the luminescence appears to emanate from centers that are located at the interface between the nanoparticles and the host. Kanemitsu[41] suggests that the interface in these cases acts as a confinement center for excitons, which are generated within the nanocrystals and combine in the interface to produce the luminescence. If this suggestion has any validity then it opens up a new field of research – the effects of interface parameters, such as interface segregants, on embedded nanoparticle associated luminescence.

In quantum dot and wire structures where the interfaces are between crystals having the same structure and nearly the same lattice parameter, the luminescence originates in the quantum wells and the interfaces contain most of the non-radiative recombination centers that reduce the efficiency of the luminescence. It has been found that the luminescent efficiency can decrease with decrease in size of the quantum well.[46] An explanation of this result is that for those quantum wells where the diffusion length is less than the smallest dimension of the quantum well a decrease of the quantum well dimension to a value less than the diffusion length will result in a greater fraction of non-radiative combination at the interface of the quantum well with its confining structure. Thus, for quantum dots and wires where the density of non-radiative recombination centers along the side wall interfaces can be high the latter interfaces can exert a controlling effect on the luminescent efficiency. Indeed, the high recombination velocity of no less than $5 \cdot 10^5$ cm/s has been measured for a 1 μm GaAs quantum dot.[46]

In luminescent polymer thin films the interface between the polymer thin films and the electrodes are sites where non-radiative recombination can occur as well as sites that can control the threshold voltage for electroluminescence through control of the overall resistance, i.e. there is an appreciable barrier to charge injection at the electrode–luminescent polymer interface. Usually, such electroluminescent polymer thin films require electron or hole or both types of injection layers on either side of the luminescent film to provide the charges that recombine within the film. It has been found that an emeraldine base (EB) polyaniline insulating film lowers the barrier to charge injection at the electrode–EB interface to near zero when placed between the electrode and the luminescent film.[60] The mechanism of this effect is not known although it has been speculated that the reduction in the interface barrier to charge injection may be due to the redox character of the EB film. The interface between electrode and luminescent polymer film has also been found to have a marked effect on the light emission efficiency for a given current. In particular, it was found that when an insulating bilayer of polymethacrylic acid (PMA)/polyallylamine hydrochloride (PAH) less than 20 Å thick was placed between an Al electrode and the light emitting poly(p-phenylene vinylene) (PPV) film the output light intensity for a given current was 2–4 times greater than in its absence.[61] This result suggests that excessive non-radiative recombination occurs at the PPV/Al interface.

3. Surface structure and properties.

3.1. Negative electron affinity.

It has been shown that hydrogen terminated (100) and (111) surfaces of natural diamond crystals have a negative electron affinity that is absent when these surfaces are no longer hydrogen terminated.[48–55] The lowering of the electron affinity is linearly related to the hydrogen coverage and a total change of 2.2 eV for hydrogen termination of a clean 2×1 (100) surface was observed and is believed due to a surface dipole created by the electropositive hydrogen.[56] On the contrary an oxygen (an electronegative atom) terminated surface increased the electron affinity by 3.7 eV. Negative electron affinity may be considered a necessary condition for a material to be used as a cold cathode in flat panel displays, ion guns, electron microscopes and the like. However, it is not a necessary and sufficient condition. The other condition which completes the sufficiency requirement is that the material be able to conduct sufficient current under low electric fields. For a display requirement the current density of $10 \, mA/cm^2$ at a field less than about $25 \, V/\mu m$ appears acceptable. It is easier to achieve a higher current density through thin films than through bulk diamond. For this reason, cold cathode schemes involving diamond films are being investigated. The conductivity of diamond films, whether undoped or p-doped, depends upon the defect concentration in the diamond, as indicated by the width of the diamond Raman peak at $1332/cm^{-1}$.[57] However, nitrogen doping improves the field emission properties of diamond-like films appreciably.[78] Further, diamond powders of diameter equivalent to film thickness, yet with sharper tips, have after atomic hydrogen treatment higher field emission current densities than the CVD diamond films at a given electric field. Other schemes of providing conducting particulates yielding high fields at their tips and embedded in a diamond-like film that has an atomic hydrogen passivated diamond-like surface also yield current densities at electric fields that are useful for flat panel applications.[58] The most effective scheme to date seems to be one of using nanometer-size diamond particles coated on a silicon substrate, which is heat treated subsequently in a hydrogen plasma at 650°C. Electron emission at applied fields of about $1 \, V/\mu m$ at an average current of $10 \, mA/cm^2$ has been reported.[77] It has been suggested that the high defect density in the nano-size particles contributes to the excellent emission properties. The technology of making flat panel displays using this technology is relatively inexpensive so that it appears likely that flat panel displays making use of these materials will become available.

3.2. Gas sensing.

There are thin film semiconductor gas sensors for which either gas adsorption on the surface or a reaction with gas adsorbed on the surface changes the number of mobile charges in the semiconductor film and hence alters its conductivity. The adsorption of an electronegative atom or molecule will tend to withdraw electrons from the conduction band and vice versa for an electropositive adsorbed species. However, the ability of the adsorbed molecules to trap or donate charge may depend upon which of the different atoms of the semiconductor they contact. For example, in an earlier section we noted that the O^{2-} ions in ZnO are relatively inert to adsorbed Pt atoms whereas the Zn^{2+} ions transfer charge to the Pt atoms and that the surface can contain different concentrations of these constituents than the bulk. This is one example of the effect of surface structure on the gas sensing phenomenon. For this or an equivalent example, saturation of the effect on conductivity will occur when there is either complete monolayer coverage by an adsorbed species or complete removal of the reacting constituent on the exposed semiconductor surface.

Other aspects of structure affect the sensitivity of the semiconductor-conductivity type gas sensor. For example, the effective surface area of the thin film can vary greatly by varying the substrate temperature during deposition and thereby controlling both the grain size and the film density, i.e. below T_1 film growth occurs to yield intergranular voiding (see Volume I, *Materials Science in Microelectronics*). The larger the surface area the greater tends to be the effect on conductivity. However, in some semiconductors, increasing the number of grain boundary barriers to charge flow will tend to decrease the charge mobility and the conductivity – an effect which opposes that of increase in the surface area. Further, in sensors dependent upon an invariant dependence of charge carrier mobility upon grain boundary barrier height, and in which the latter parameter is varied by control of charge carrier density via adsorption of donors or acceptors, it is found that grain boundary diffusion of species driven by exposure to oxidizing or reducing gases can alter the functional dependence between conductivity and surface adsorbed species concentration, i.e. introduce a sensor instability.[62] This instability can only be combatted by making the grain boundary traps insensitive to oxidation/reduction.

Another effect of structure on the semiconductor-conductivity type of gas sensor is provided by the presence of additional phases along the surface of the semiconductor. How these additional phases exert their effect may not be known in many cases. For example, and again for a ZnO film, it was found that the conductivity increased more than 50-fold when the film's atmosphere was changed from air to 200 ppm in air of trimethylamine, a gas given off when seafood deteriorates, and that this effect could be multiplied by a factor of three by doping the ZnO with 5 wt% Al_2O_3.[63] In another example, the presence of Pt and a rare earth oxide phase in contact with a SnO_2 surface enhanced the sensitivity of the SnO_2 semiconductor to CO detection compared to that obtained on a bare SnO_2 surface.[64]

Another class of sensors is based on a diode structure in which there is a thin insulator film between the semiconductor and metal film such that equalization of the Fermi levels of semiconductor and metal is prevented. This diode operates as a hydrogen sensor when the metal is palladium and the sensing is accomplished as a consequence of the effect of hydrogen in changing the work function of the palladium film. The insulator film thickness is sufficiently small so that tunneling of electrons through it is not the factor limiting the current but rather it is the barrier corresponding to the difference between the semiconductor conduction band level and that of the metal that limits the current. As the metal work function changes so will this activation energy barrier. Structure affects the operation of this diode through its effect on the insulator film in preventing the formation of a metal silicide, as a consequence of metal atom diffusion through the insulator film, or an oxide on the semiconductor surface. Since the insulator must be thin to ensure tunneling care must be taken to produce an integral, diffusion barrier grade, oxide film.[71] Another Schottky barrier thin film hydrogen sensor consists of a Pd film directly on a TiO_2 semiconductor film without an intervening insulator film.[72] Although the mechanism involving the sensitivity to hydrogen is questionable, if hydrogen affects the work function of Pd, as suggested above, then it may affect the Schottky barrier height, if the junction is an ideal one, or if the junction is not ideal, the effect on SBH may be that of a dipole layer, i.e. see Figure 6.3.

There are still other effects of adsorbed gases that enable them to be sensed. For example, the ability of current to percolate through grain boundaries of a film, produced by sputtering from a target consisting of 90%W and 10%Ti onto Si(110) at 300°C in an $Ar-O_2$ mixture at a pressure of $4 \cdot 10^{-3}$ mbar and then annealed for 12 hours at 600°C, is affected by the presence of NO_2. This ability has been used to produce a sensor for NO_2.[73] The sensitivity of this sensor increases as the grain size decreases. Similar effects of grain size on gas sensing sensitivity have been found by others.[74] For these effects of grain size to be detected the gases involved must somehow affect the ability of current to percolate through the film. The obvious candidate for such an effect is one in which the gases involved can penetrate the intergranular space and directly affect the intergranular resistance. We have already remarked upon the fact that films deposited with substrates at a temperature in zone 1 will have intergranular voids and will be permeable to gases and other molecules (see Volume I, *Materials Science in Microelectronics*). Indeed, it has been shown that Cl will penetrate intergranular voids in films of copper[75] deposited in zone I and will not do so in films deposited under conditions where there are no intergranular voids.[76]

3.3. Catalysis.

Catalysts require a very high surface area per unit volume. Thus, catalysts are not usually in the form of thin films, but rather in the form of very fine particles. However, a strategy has been proposed for preparing controlled porosity ceramic thin films having high surface area per unit volume using sol–gel techniques.[65] It is not known whether such films of catalytic materials have been prepared and tested as yet. Should this strategy be successful then the resulting thin films would represent an interesting example of the effect of structure in thin films on catalysis.

3.4. Corrosion.

Since surfaces are present on both thin film and bulk forms of a given material corrosion may be expected for both geometries. The question is whether the rate of corrosion will be different or not. The answer to this question is that the rate of corrosion is not dependent on the dimension of the object but rather upon the degree of inhomogeneity in structure and composition present at the surface under exposure to water vapor or other corrosive media. Some thin films have compositions that may be especially prone to corrosion because they are not homogeneous in either composition or structure. For example, some CoCr films used in magnetic recording application consist of a mixture of Cr rich and Cr poor areas (see Chapter II, 2.2.1). Thus, there are concentration cells at the surface of this film, which in the presence of water vapor, would provide the driving force for corrosion of the film. The degree of heterogeneity of these cells can be varied by processing conditions. Hence, it is no surprise that it was found that the thermal activation energy for the corrosion of these films varied from 0.07 to 0.3 eV with variation of the sputtering deposition conditions.[66] Usually, the optimum microstructure for corrosion resistance does not correspond to the optimum one for the properties controlling the choice of the film, e.g. the CoCr film is used for its magnetic recording properties. Hence, other means of corrosion protection of these films must be used.

From the point of view of structure and its effect on corrosion any overcoat film used for corrosion protection must be homogeneous and not contain pinholes or other passages that allow contact between the environment and the substrate being protected. Thus, protective film layers should not be deposited at temperatures below T_1 (see Volume I, *Materials Science in Microelectronics*, Chapter I) unless a deposition technique is used that yields absolutely zero percent porosity. What types of films provide corrosion protection? The answer is non-porous films, which are either noble metals or metals which passivate (alone or alloyed). Also, films which are galvanically compatible with their substrates are feasible as corrosion protective films.

Recapitulation

Ideal Schottky junctions are not found in metal–semiconductor systems because the charge transfer between these components is a function not only of the electronegativity difference, but also of the semiconductor's electronic susceptibility. Further, the SBH is sensitive to a host of interface parameters, including structure. The contact resistance of ohmic junctions can be affected by doping the zone of the Schottky junction. Defects, intermediate phases, anion segregation all act to increase the contact resistance of ohmic junctions. The interface defect state density at Si/SiO_2 interfaces depends upon many variables including processing parameters. It appears to be related to the step density. The total interface defect density consists of P_b dangling bonds including such dangling bonds that are passivated by hydrogen. The latter are the main source of deleterious effects on properties in that they lead to multiplication of the dangling bond density.

Although it has been difficult to produce a low enough defect state density at insulator–compound semiconductor interfaces, this has been finally accomplished using a technique that minimizes the density of antisite defects at the surface. Such a technique is also useful in producing a passivation of the compound surface when there is a continuity of bonding across the interface between the compound semiconductor and its contacting layer.

Non-radiative recombination center free interfaces can act as radiative recombination centers, perhaps by acting as confinement centers for excitons that are generated within the nanocrystals these interfaces bound. However, most interfaces contain non-radiative recombination centers which act to limit the luminescence from quantum dots, wires, etc. An appropriate and thin (enough to allow electron tunneling through it) insulating polymer film at the interface between electroluminescent polymer films and metal electrodes can prevent the non-radiative recombination at this interface and allow radiative recombination to occur elsewhere.

Diamond films have a negative electron affinity when their surfaces are passivated by hydrogen. The use of such films as thin layers over point field emitters offers a possibility for their use in flat panel displays.

Thin film sensors take advantage of several effects of gas adsorption: one is to reduce the number of majority charge carriers by trapping, another is to affect the work function of palladium (mainly for gases containing hydrogen as a constituent) and another is to provide enhanced intergranular contributions to the resistance in films that have intergranular void networks.

References

1. J.C. Phillips, **Bonds and Bands in Semiconductors**, Academic Press, New York, 1973, p. 254.
2. J. Tersoff, Phys. Rev. Lett. <u>52</u>, 465(1984); Phys. Rev. <u>B30</u>, 4874(1984).

3. V. Heine, Phys. Rev. A138, 1689(1965).
4. M. Lannoo and P. Friedel, **Atomic and Electronic Structure of Surfaces**, Theoretical Foundations, Springer-Verlag, 1991.
5. M. Cardona and N.E. Christensen, Phys. Rev. B35, 6182(1987); and Refs. 2 and 4.
6. W. Mönch, **Semiconductor Surfaces and Interfaces**, Springer, 1995.
7. A.R. Miedema, F.R. de Boer and P.F. de Chatel, J. Phys. F (Metal Physics). 3, 1558(1973); A.R. Miedema, P.F. de Chatel and F.R. de Boer, Physica. B100, 1(1980).
8. W.T. Petrie and J.M. Vohs, MRS Symp. Proc. 357, 17(1995).
9. W. Mönch, J. Appl. Phys. 80, 5076(1996).
10. R. Tung, in **Contacts to Semiconductors**, ed. L.J. Brillson, Noyes Publications, Park Ridge, NJ, 1993.
11. J.R. Waldrop, J. Vac. Sci. Tech. B3, 1197(1985); Appl. Phys. Lett. 47, 1301(1985).
12. Ref. 6, p. 370.
13. R.T. Tung, J.P. Sullivan and F. Schrey, MRS Symp. Proc. 318, 3(1994).
14. J.P. Sullivan, W.R. Graham, F. Schrey, D.J. Eaglesham, R. Kola and R.T. Tung, MRS Symp. Proc. 320, 249(1994).
15. P. Werner, W. Jäger and A. Schüppen, ibid., p.227.
16. K.L. Kavanagh, B.A. Morgan, A.A. Talin, K.M. Ring, R.S. Williams, M.C. Reuter and R.M. Tromp, MRS Symp. Proc. 402, 461(1996).
17. R.M. Anderson and T.M. Reith, J. Electrochem. Soc. 122, 1337(1975); H.B. Ghozlene and P. Beaufrere, J. Appl. Phys. 49, 3998(1978).
18. P.W. Pellegrini, C.E. Ludington and M.M. Weeks, J. Appl. Phys. 67, 1417(1990).
19. I. Ohdomari, T.S. Kuan and K.N. Tu, J. Appl. Phys. 50, 7020(1979); Phys. Rev. B42, 5249(1990).
20. Q.F. Wang, A. Lauwers, F. Jonckx, M. de Potter, C.-C. Chen, and K. Maex, MRS Symp. Proc. 402, 221(1996).
21. P.W. Leech and G.K. Reeves, MRS Symp. Proc. 318, 183(1994).
22. S. Oktyabrsky, M.O. Aboelfotoh and J. Narayan, MRS Symp. Proc. 402, 541(1996).
23. N.S. Fatemi and V.G. Weizer, 318, 171(1994).
24. G. Lucovsky, J.T. Fitch, E. Kobeda and E.A. Irene, in **The Physics and Chemistry of SiO2 and the Si-SiO2 Interface**, eds. C.R. Helms and B.E. Deal, Plenum Press, New York, 1988, p. 139.
25. J.E. Olsen and F. Shimura, J. Appl. Phys. 66, 1353(1989).
26. I.W. Boyd and I.B. Wilson, J. Appl. Phys. 62, 3195(1987).
27. F.J. Grunthaner, P.J. Grunthaner, R.P. Vasqueg, B.F. Lewis and J. Maserjian, Phys. Rev. Lett. 43, 1683(1979).
28. T. Hattori and T. Suzuki, Appl. Phys. Lett. 43, 470(1983).
29. D.J. DiMaria, Microelectron. Eng. 28, 63(1995).
30. N. Bhat and K.C. Saraswat, in **Silicon Nitride and Silicon Dioxide Thin Insulating Films**, eds. V.J. Kapoor and W.D. Brown, Proceedings, Vol. 94-16, The Electro Chemical Society, Pennington, NJ, 1994, p. 23.
31. G. Lucovsky, D.R. Lee, Z. Jing, J.L. Whitten, C. Parker and J.R. Hauser, MRS Symp. Proc. 405, 321(1996).
32. M. Passlack and M. Hong, MRS Symp. Proc. 421, 81(1996).
33. J.M. Woodall and J.L. Freeouf, J. Vac. Sci. Technol. 19, 794(1981).

34. A. Thilderkvist, J. Michel, S.-T. Ngiam, L.C. Kimerling and K.D. Kolenbrander, MRS Symp. Proc. 405, 265(1996).
35. K.S. Min, K.V. Shcheglov, C.M. Yang, R.P. Camata, H.A. Atwater, M.L. Brongersma and A. Polman, ibid, p. 247.
36. Y. Maeda, N. Tsukamoto, Y. Yazawa, Y. Kanemitsu and Y. Masumoto, Appl. Phys. Lett. 59, 3168(1992).
37. Y. Kanemitsu, T. Ogawa, K. Shiraishi and K. Takeda, Phys. Rev. B48, 4883(1993).
38. H. Takagi, H. Ogawa, Y. Yamazaki, A. Ishizaki and T. Nakagiri, Appl. Phys. Lett. 56, 2349(1990).
39. F. Koch, V. Petrova-Koch, T. Muschik, A. Nikolov and V. Gavrilenko, MRS Symp. Proc.283, 197(1993).
40. T. Matsumoto, J. Takahashi, T. Tamaki, T. Futagi, H. Mimura and Y. Kanemitsu, Appl. Phys. Lett. 64, 226(1994).
41. Y. Kanemitsu, MRS Symp. Proc. 405, 229(1996).
42. M. Oshima, T. Scimeca, Y. Watanabe, H. Oigawa and Y. Nannichi, Jpn. J. Appl. Phys. 32, 518(1993).
43. Y.-L. Chang, I.-H. Tan, Y.-H. Zhang, J. Merz, E. Hu, A. Frova and V. Emiliani, Appl. Phys. Lett. 62, 2697(1993); Y.-I. Chang, S. I Yi, S. Shi, E.L. Hu, W.H. Weinberg and J.L. Merz, J. Vac. Sci. Technol. B13, 1801(1995).
44. T. Hattori and K. Ohishi, MRS Symp. Proc. 318, 61(1994).
45. S.S. Shi, Y.-I. Chang, E.L. Hu and J.J. Brown, MRS Symp. Proc. 421, 401(1996).
46. L.-L. Chao, M.B. Freiler, M. Levy, J.-L. Lin, C.S. Cargill III, R.M. Osgood Jr. and G.F. McLane, MRS Symp. Proc. 405, 429(1996).
47. Z. Jing, G. Lucovsky and J.L. Whitten, MRS Symp. Proc. 318, 287(1994).
48. F.J. Himpsel, J.A. Knapp, J.A. van Vechten and D.E. Eastman, Phys. Rev. B20, 624(1979).
49. B.B. Pate, M.H. Hecht, C. Binns, I. Lindau and W.E. Spicer, J. Vac. Sci. Technol. 21, 268(1982).
50. J. van der Weide and R.J. Nemanich, Appl. Phys. Lett. 62, 1878(1993).
51. N. Eimori, Y. Mori, A. Hatta, T. Ito, and A. Hiraki, Jpn. J. Appl. Phys. 33, 6312(1994).
52. D.P. Malta, J.B. Posthill, T.P. Humphreys, R.E. Thomas, G.G. Fountain, R.A. Rudder, G.C. Hudson, M.J. Mantini, and R.J. Markunas, Appl. Phys. Lett. 64, 1929(1994).
53. J. van der Weide, Z. Zhang, P.J.K. Baumann, M.G. Wensell, J. Bernholc and R.J. Nemanich, Phys. Rev. B50, 5803(1994).
54. P.K. Baumann and R.J. Nemanich, Diam Relat Mater. 4, 802(1995).
55. C. Bandis and B.B. Pate, Phys. Rev. Lett. 74, 777(1995).
56. R.E. Thomas, T.P. Humphreys, C. Pettenkofer, D.P. Malta, J.B. Posthill, M.J. Mantini, R.A. Rudder, G.C. Hudson and R.J. Markunas, MRS Symp. Proc. 416, 263(1996).
57. W. Zhu, G.P. Kochanski and S. Jin, ibid. p. 443.
58. H. Sohn, K. Krishnan and R. Fink, ibid. p. 455.
59. A. Stesmans and G. van Gorp, Appl. Phys. Lett. 57, 2663(1990).
60. A.G. Macdiarmid, H.L. Wang, F. Huang, J.K. Avlyanov, P.C. Wang, T.M. Swager, Z. Huang, A.J. Epstein, Y. Wang, D.D. Gebler, R. Shashidhar, J.M. Calvert, R.J. Crawford, T.G. Vargo, G.M. Whitesides, Y. Xia, and B.R. Hsieh, MRS Symp. Proc. 413, 3(1996).
61. M. Ferreira, O. Onitsuka, A.C. Fou, B.R. Hsieh and M.F. Rubner, ibid. p. 49.

62. D. Narducci, S. Pizzini and F. Morazzoni, in **Proceedings of the Symposium on Chemical Sensors**, Vol. 93-7, eds. M. Butler, A. Ricco and N. Yamazoe, The Electrochemical Society, Pennington, NJ, 1993, p.325.

63. H. Nanto, T. Kawai and S. Tsubakino, ibid. p. 522.

64. D. Narducci, L. Cinquegrani, E. D'Acci and S. Pizzini, ibid. p. 482.

65. C.J. Brinker, R. Sehgal, N.K. Raman, S.S. Prakash and L. Delattre, MRS Symp. Proc. 368, 329(1995).

66. K. Tagami and H. Hayashida, IEEE Trans. Magn. Mag-23, 3648(1987).

67. E.H. Poindexter, G.J. Gerardi, M.-E. Rueckel, P.J. Caplan, N.M. Johnson and D.K. Biegelsen, J. Appl. Phys. 56, 2844(1984); G.J. Gerardi, E.H. Poindexter, P.J. Caplan and N.M. Johnson, Appl. Phys. Lett. 49, 348(1986); J.H. Stathis and L. Dori, Appl. Phys. Lett. 58, 1641(1991).

68. N.S. Saks and D.B. Brown, IEEE Tran. Nucl. Sci. NS-36, 1848(1989); R. Stahlbush, B. Mrstik and R. Lawrence, IEEE Tran. Nucl. Sci. NS-37, 1641(1990); S.J. Wang, J.M. Sung and S.A. Lyon, Appl. Phys. Lett. 52, 1431(1988); P.S. Winokur, H.E. Boesch, J.M. McGarrity and F.B. McLean, J. Appl. Phys. 50, 3492(1979).

69. V.J. Kapoor and M.Shokrani, in **Silicon Nitride and Silicon Dioxide Thin Insulating Films**, eds. V.J. Kapoor and W.D. Brown, Proceedings, Vol. 94-16, The Electro-chemical Society, Pennington, NJ, 1994, p. 95.

70. T. Kobayashi, T. Ichikawa and T. Sawai, Appl. Phys. Lett. 49, 351(1986).

71. R.C. Hughes, W.K. Schubert, T.E. Zipperian, J.L. Rodriguez and T.A. Plut, J. Appl. Phys. 62, 1074(1987).

73. M. Ferroni, V. Guidi, G. Martinelli, and G. Sberveglieri, J. Mater. Res. 12, 793(1997).

74. H. Pink, L. Treitinger and L. Vite, Jpn. J. Appl. Phys. 19, 513(1980).

75. E.S. Machlin and Yu-Ren Zhang, Mat. Sci. Eng. B3, 335(1989).

76. T. Park, T.N. Rhodin and L.C. Rathbun, J. Vac. Sci. Tech. A4, 168(1986).

77. W. Zhu, G.P. Kochanski and S. Jin, MRS Symp. Proc. 498(1998).

78. G. Amaratunga, S.R.P. Silva, Appl. Phys. Lett. 68, 2529(1996); B.S. Satyaraman, A. Hart, W. Meline and J. Robertson, submitted to Appl. Phys. Lett. (1997).

79. G.P. Das, P. Blöchl, O.K. Andersen, N.E. Christensen and O. Gunnarsson, Phys. Rev. Lett. 41, 1168(1989); H. Fujitani and S. Asano, Phys. Rev. B42, 1696(1990).

80. S. Chang, L.J. Brillson, D.S. Rioux, S. Kirchner, D. Pettit and J.M. Woodall, J. Vac. Sci. Tech. B8, 1008(1990); Phys. Rev. B44, 1391(1991).

81. C.J. Palmstrom, T.L. Cheeks, H.L. Gilchrist, J.G. Zhu, C.B. Carter and R.E. Nahory, MRS Ext. Abstr., EA-21, 63(1990); C.J. Palmstrom, T.L. Cheeks, H.L. Gilchrist, J.G. Zhu, C.B. Carter, B.J. Wilkens and R. Martin, J. Vac. Sci. Tech. A10, 1946(1992).

82. J.M. Woodall, G.D. Pettit, T.N. Jackson, C. Lanza, K.L. Kavanaugh and J.W. Mayer, Phys. Rev. Lett. 51, 1783(1983).

Bibliography

1. T. Hattori, Chemical Structures of the SiO_2/Si Interface, Crit. Rev. Solid State Mat. Sci. 20, 339(1995), CRC press.

Appendix

One of the significant interface developments in thin films encompasses the Si/high K dielectric MOS devices. This MOS device is characterized by an unacceptable threshold voltage. The origin of this threshold voltage is not yet established. However, it is known that a dipole exists across a HfO_2 layer, that the Fermi level is pinned in the upper part of the band gap at a p-poly Si/HfO_2 interface.[A1] It is believed that a Hf–Si reaction at the interface between the HfO_2 and poly Si is responsible for the trap states in the upper band gap of Si that cause the Fermi level pinning. A reaction product, perhaps $HfSiO_4$, has been found at the Si/HfO_2 interface.[A2] This product increases upon a 400°C anneal[A2] as does the number of disorder induced gap states[A3] while the number of interface trap states decreases. Although the latter anneal was carried out in forming gas, a rapid initial transfer of oxygen from the HfO_2 might still enable the formation of a reaction product and an increase in the number of disorder induced gap states. A possible solution to the reaction problem is to place a thin Al_2O_3 capping layer between the HfO_2 and poly-Si.[A4] Although this solution results in a 10^4 decrease in leakage current[A5] it is not known whether the Al_2O_3 capping layer is also a solution to the threshold voltage problem since a threshold voltage shift has been found for Al_2O_3 gate dielectric MOS devices.[A6] Since according to Ludeke et al.[A1] this shift is associated with the formation of a dipole across the dielectric layer one may speculate that the dipole arises from the transport of charge across the dielectric layer induced by a reaction on one side of this layer. Thus, a solution may involve elimination of asymmetry in chemical reactions at the interfaces of the dielectric.

The large surface/volume ratio of nanotubes enhance their suitability over other morphologies in sensor applications. Such sensors have reached the commercial stage and have numerous applications as a search via Google will reveal to the curious reader.

References to Appendix

A1. R. Ludeke et al., IBM Research Report RC23474 (W0412-066) December 21, 2004, Physics.
A2. A. Modreanua et al., Optical characterization of high-k dielectrics HfO_2 thin films obtained by MOCVD, Request report from mircea.modreanu@nmrc.ie.
A3. S. Duenas et al., MRS Symp. Proc. 786, E3.18.1(2004).
A4. M. Kundu et al., Appl. Phys. Lett. 82, 3442(2003).
A5. D.C. Gilmer et al., Appl. Phys. Lett. 81, 1288(2002).
A6. J. H. Lee et al., Tech. Dig. – Int. Electr. Dev. Meet. 2000, 645(2000).

Problem

1. What observation precludes the possibility that ionicity induced charge transfer dipole formation can be responsible for the non-ideal values of Schottky barrier heights measured in compound semiconductors?
2. Where would the defect level be relative to the conduction band minimum for the case that the Schottky barrier height is (a) 0, (b) 0.5 eV?
3. When only two Schottky barrier heights are observed for a given semiconductor independent of the metal at the junction, what must this mean in terms of defect pinning of the Fermi level?
4. If dangling bonds at Si/SiO$_2$ interfaces are so deleterious to properties of MOS devices why is their passivation by hydrogen not a solution to this problem?
5. What properties will passivation of dangling bonds at interfaces improve?
6. What would happen to the electron affinity of a diamond film if it were heated in a vacuum at elevated temperature?
7. Why should the current emitted from a diamond film depend upon the morphology of a conducting layer surface beneath it?
8. How does gas adsorption act to affect the work function of a metal?
9. Would you expect that a thin film semiconductor doped with acceptor impurities would be sensitive to adsorbed electropositive gases?
10. Which deposition procedure would you use to produce a thin film sensor that depends upon sensing the intergranular resistance change due to adsorbed gases?

CHAPTER VII

Defects and Properties

The objective of this chapter is to provide additional knowledge about defects and their effects on properties beyond that already given in previous chapters. Also, the section headings in this chapter will deviate from the formula used in the previous chapters. In this chapter the section headings are provided by the structural units rather than by properties. Hopefully, this new approach will provide new understanding of the relations between defects and properties.

1. Grain boundaries.

1.1. Semiconductors.

Grain boundaries affect a myriad of electrical and other properties in semiconductors. In the following we shall consider first the structure of grain boundaries in tetrahedrally bonded semiconductors and the possible defects that these grain boundaries may encompass. Then we will consider those effects of grain boundaries and their associated defects on properties which have not already been discussed in the other chapters of this book.

1.1.1. Structure.

To the best of this writer's knowledge the only theoretical studies on grain boundaries in silicon reported in the literature prior to 2004 involve tilt and twist boundaries only,[47] except for one preliminary study of a random high angle boundary.[52] These theoretical studies[47] of tilt grain boundaries in silicon suggest a multitude of modes of bonding all involving fourfold coordination. The simplest boundaries studied have a perfect <011> tilt axis and contain the tilt axis. The calculated energies of such tilt boundaries are shown in Figure 7.1. There are but two defects in such boundaries: Lomer dislocations and twin units. The $\Sigma 9$ boundary was shown to have Lomer dislocations (a/2[011]) with a zigzag arrangement of five and seven membered rings that make up the cores of these dislocations. There are no states in the energy gap provided by this grain boundary. Small angle tilt boundaries contain straight arrays of Lomer dislocations* with larger angle boundaries having

* A view down the core of a straight Lomer dislocation revealing 5 and 7 member rings is shown in Appendix 1.

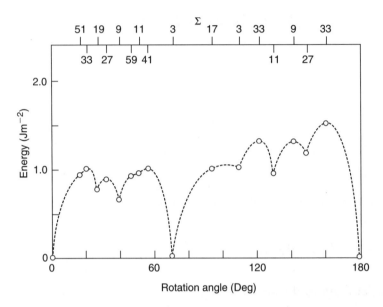

Figure 7.1. Calculated energy versus rotation angle about <110> tilt axis for grain boundaries in silicon. (Reproduced with permission from M. Kohyama, Model. Simul. Mater. Sci. Eng. <u>10</u>, R31(2002). Copyright IOP Publishing Ltd.)

denser spacing of these cores. At the Σ9 boundary the zigzag arrangement is introduced. In iso-lated Lomer dislocations the bonds are strained more than in the Σ9 boundary and may introduce tail states into the energy gap.

The most stable grain boundary in this class of <011> tilt boundaries is the Σ3 (111) twin at the tilt angle of 70.5°. This boundary consists of boat shaped 6 member rings. For higher tilt angles the boundary structure becomes more complicated involving reconstruction of dangling bonds. Nevertheless, there are no deep or shallow states in the energy gap for these boundaries. (Boundaries having the <011> tilt axis but not perfectly aligned with the tilt axis probably contain secondary dislocations that are electrically active.)

Similarly, boundaries, that are perfectly aligned along a <100> tilt axis, contain dislocation cores in which the atoms are all 4 coordinated and do not contribute states within the energy gap. Indeed, all tilt type coincidence boundaries that have been investigated theoretically in elemental semiconductors have no dangling bonds and do not contribute states within the energy gap. Experimental studies[48] verify the theoretical results just described.

However, should there be a twist component in the grain boundary then other than 4-coordinated sites can be present corresponding to dangling bonds with gap states localized about the coordination defects.[47,52] This statement is valid for twist-type coincidence boundaries also.

Reconstructions in Si and Ge yield 4-coordination arrangements in grain boundaries that have been disturbed by vacancy condensation and the like. However, this reconstruction in diamond does not yield lower energy configurations than those with coordination defects. The higher stiffness of the diamond bonds are believed to be responsible for this situation.

Grain boundaries in III–V and IV–IV compounds (GaAs, SiC) having fourfold coordination, such as in the sphalerite and wurtzite structure, are believed to behave very much as they do for Si. That is, the atoms in tilt boundaries are fourfold coordinated, whereas high angle boundaries contain coordination defects. However, in compound semiconductors there is the possibility of having wrong bonds between like atoms in grain boundaries. For certain boundaries these wrong bonds do not give rise to dangling bonds. For example, in the case of $\Sigma 19$ [110] boundaries in GaAs there are Ga–Ga bonds but no dangling bonds.[53]

Grain boundaries in cubic SiC have also been studied theoretically.[47] These results suggest that polar boundaries may be stable relative to the non-polar one for the $\Sigma 9$ {122} coincidence boundary. However, experiment results in the observation of the non-polar boundary. It is suggested that this was brought about because the $\Sigma 9$ boundary was at a triple junction with a $\Sigma 3$ boundary that prefers the non-polar one energetically. Apparently, more study of the grain boundaries in SiC is needed.

In the semiconductors CdTe and CuInSe$_2$ each atom is fourfold coordinated as well. Thus, without evidence to the contrary it seems reasonable to suppose that the grain boundaries in these structures behave similarly to Si and the III–V compounds in the sense that tilt boundaries have no dangling bonds but that high angle boundaries and twist boundaries do. The double-positioning twin boundaries in CdTe, however, do have associated dangling bonds as well[69] and introduce states into the gap.

Contrary to prior theoretical calculations Kim et al.[71] have shown both theoretically and experimentally that grain boundaries in SrTiO$_3$ are intrinsically non-stoichiometric. There tends to be an excess of Ti over O in the grain boundaries. This is the factor that yields charged grain boundaries in SrTiO$_3$. It is unusual in that the charge comes about intrinsically and not by the usual means of impurity segregation.

From the above summary of our knowledge concerning structure of grain boundaries in semiconductors it is apparent that much remains to be learned in this regard for the compound semiconductors of the III–V and II–VI classes.

1.1.2. Electrical properties.

1.1.2.1. Origin and effects of grain boundary potential barriers.

We have already alluded to two sources of grain boundary potential barriers: charged impurity atoms and non-stoichiometry in the grain boundaries of compounds. Another origin in $Cu(In,Ga)Se_2$, and perhaps in other compounds, has been suggested by the work of Perssons and Zunger.[54] It is based on a first principle calculation of the state of a grain boundary in the $CuInSe_2$ system. A result of this calculation is that "local reconstruction at the grain boundary expels holes, thus creating a 'free zone' for fast electron transport." It accomplishes this feat via d–p interactions resulting from Cu vacancies in the grain boundary which are *electrostatically stable* via a complex of the form $(2V_{Cu}^- + In_{Cu}^{++})^\circ$. The significance of a "free zone", say of holes, is that electrons will not be impeded in passing through the boundary since trapping and recombination is impossible with only one sign of charge and that the conductivity along the grain boundary will be enhanced since again there is no trapping or recombination in this region. The grain boundary is *not* electrically charged. However, a barrier to the transport of holes into the boundary will be developed.

Jiang et al.[95] have found a built-in potential at grain boundaries in $Cu(In,Ga)Se_2$ the maximum value of which correlates to the dependence of solar cell efficiency on the Ga/Ga + In ratio. Thus, in this case the grain boundaries have the properties predicted by Persson and Zunger,[54] although there is no proof that the grain boundary potential is a consequence of their model. Nevertheless, the experimental evidence that there are grain boundaries in compound semiconductors that have the properties of collecting photo generated charge and *delivering the charge to electrodes they intersect* is a new development which will undoubtedly spark new research. It is the delivery part that is new and much needs to be learned as to what aspects of the microscopic structure of the grain boundaries allows the enhanced sheet conductance for the minority carrier. The band structure calculations of Persson and Zunger are welcome because they show how grain boundaries can develop a potential barrier without an origin in electrical charge along grain boundaries, at least in $CuInSe_2$ and thereby achieve a state of enhanced conductivity for the charge attracted to it by the potential.

Moller[7] has shown that the presence of Cu vacancies at grain boundaries in $CuInSe_2$ is reasonable. In particular, the number of like atom nearest-neighbor contacts at grain boundaries is high. These contacts would lead to high energy configurations if atoms occupied the sites corresponding to these contacts while vacancies at the contacting sites would yield a lower energy configuration. Hence, the possibility of vacancies at one of the contacting sites is reasonable.

Normally, grain boundary barriers form because of electrical charges at grain boundaries. These charges are either majority carriers trapped there or charged impurities. These barriers prevent the traversal of the boundary by both minority and majority charge carriers for different reasons. The charge carriers having the same charge as the impurities are repelled from the boundary. Those having the opposite charge are attracted to the boundary where they are either trapped or recombine with trapped opposite sign carriers and can only succeed in reaching the other side under the influence of a bias voltage either by thermionic emission from traps or by tunnelling through the barrier about a trap. A typical barrier of this kind is illustrated in Figure 7.2 for the case where there is a bias voltage across the barrier.

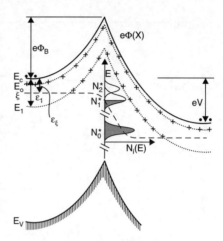

Figure 7.2. Schematic illustration of barrier at grain boundary. Inset shows trap states in energy gap. Fermi energy level is given by dashed line.

The grain boundary barrier of Persson and Zunger has characteristics that differ markedly from the normal one. Although it acts to repel holes in p-$Cu(In,Ga)Se_2$ it is transparent to electrons. No traps are in the boundary to prevent electron transport either through or along the boundary. No holes are trapped in the boundary to recombine with the electrons. The boundary region and its associated depletion zones are regions in which the effective diffusion length is larger than in the bulk crystal since holes and electrons are separated in such a way that these regions only contain electrons, i.e. no recombination occurs in these regions.

Grain boundaries along which charge inversion has occurred have different barrier properties also. Charge inversion is brought about when the barrier potential exceeds the energy gap. The potential height is affected by various parameters. One is the density of interface (grain boundary) states in the energy gap. An increase in this density increases the height of the grain boundary barrier potential and thus can induce charge inversion at the boundary. The barrier (and its sign) remains in the presence of inversion, although it becomes insensitive to the density of interface states. These qualities are revealed in Figure 7.3. The quantity N_{itc} is the critical value of the interface state density at which inversion occurs. One sees from this figure that the barrier potential maintains the same sign but levels off after inversion. These are theoretical results of Chattopadhyay.[64]

Among the different properties of inversion boundaries is the existence of enhanced electrical sheet conductance. The free minority charge carriers in the inversion boundary are able to move freely along the boundary. Inversion boundaries have been found in InSb, $Hg_{1-x}Cd_xTe$ and Ge.[57] The boundaries in these two systems may be described as follows: (1) A degenerate n-type inversion layer exists adjacent to the grain boundary in p-type bicrystals. (2) The quasi-2D-character of the electron gas confined in the symmetrical quantum well of the grain boundary interface in narrow-gap semiconductors is well established. (3) Calculations indicate the existence of two types of ionized grain boundary states: positively charged donor states localized in the grain boundary core with a density of $4 \cdot 10^{12} cm^{-2}$

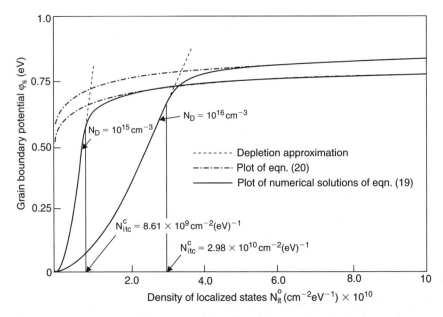

Figure 7.3. Calculated grain boundary barrier potential versus density of states at the grain boundary. (Reproduced with permission from P. Chattopadhyay, J. Phys. Chem. Sol. **56**, 189. Copyright 1995 Elsevier.)

and negatively charged ones more spatially extended with a density of $2.3 \cdot 10^{12} \, \mathrm{cm}^{-2}$. It appears as if the positive charged core states are screened by a more spatially extended negative charge region so that the electron gas in the quantum well adjacent to the grain boundary is unaffected by the positive core.

Another property that can be inferred from the above described characteristics of inversion boundaries is that they do not harbor recombination processes because the potential expels majority carriers from the boundary region and the minority carriers that are drawn into the boundary are shielded from contact with the positive charges at traps along the grain boundary core region. Still another property that can be inferred is that an inversion boundary is still a barrier to the transport across the boundary of both sign of charge carriers. Hence, the expectation is for it to hinder both recombination within the boundary region and transport of charge carriers from one side to the other side of the boundary while it will not hinder transport of minority charge along its sheet.

A variety of interesting observations have been made that may be related to inversion boundaries in solar cells and YBCO. In CdTe a deep defect present in grain boundaries is a Cd vacancy (V_{Cd}) and also its impurity related complexes.

It is an acceptor and probably a non-radiative recombination center. It has been proposed that ClTe forms a defect complex with V_{Cd} (the chlorine A center) and perhaps in this way contributes to the passivation of the grain boundaries in chlorinated CdTe.[58] Chlorine passivation results in the removal of EBIC contrast at grain boundaries of CdTe[58] (i.e. non-radiative recombination in the grain boundary core is greatly reduced). The process of chlorination also greatly enhances the conductivity of the grain boundary along its interface[59] as well as produces a depletion region next to the grain boundary in p-CdTe[59] and sets up a barrier in this absorber layer to the transport of holes across grain boundaries.[59,60,61]

One aspect of this barrier is that it separates holes and electrons photo-generated in the barrier region, sending them to and away from the grain boundary core. This process has been experimentally verified for chlorinated CdTe.[67] This result has been given as the explanation for the observation that polycrystalline CdTe exhibits higher solar efficiency than single crystal CdTe. However, all grain boundaries in semiconductors that contain trap states are likely to develop band bending induced potential barriers which will separate photo generated holes and electrons. Yet single crystal efficiencies in GaAs and Si, for example, exceed those for polycrystalline films. Hence, the separation of photo-generated holes and electrons within the potential barrier layer about grain boundaries cannot by itself explain the observation that single crystal CdTe has lower solar efficiency than polycrystalline CdTe.

The fact is that unchlorinated CdTe polycrystalline film does not exhibit enhanced current along grain boundaries when subjected to the same light injection and yet it exhibits a depletion zone about grain boundaries characteristic of a band bending induced potential barrier although the depletion width is reported to be smaller (100–250 nm) in unchlorinated CdTe than in chlorinated CdTe (200–350 nm).[59] One would expect if the charge carriers photo generated in the depletion zone were to reach the electrodes the current in the grain boundaries of the unchlorinated films would be no smaller than one-half that in the chlorinated films. However, the evidence offered in the paper by Visoly-Fisher et al.[59] does not support this expectation. Rather, it supports the assertion that the photo generated charge carriers directed toward the grain boundary in the unchlorinated film do not arrive at the electrodes.

Given these facts it has been possible to speculate concerning their origin. Visoly-Fisher et al.[59] invoke inversion (band bending sufficient to bring the bottom of the conduction band down to the Fermi level) at the grain boundary to explain both the enhanced conductivity and the existence of a barrier to hole transport across the grain boundary and a hole depletion zone adjacent to the grain boundary. As noted in the above these are characteristics of inversion grain boundaries. Despite this correspondence between observed and expected behavior additional experiments are needed to prove that these boundaries are indeed inversion boundaries.

In SrTiO$_3$ tilt grain boundaries about a <001> axis have been studied experimentally and theoretically and shown to be non-stoichiometric.[71] The consequence of this state is that in p-doped $SrTiO_3$ the boundary cores become positively charged setting up adjacent space charge regions and acting like a p–n–p double junction. Thus, contrary to the statement in Reference 56 that charged grain boundaries in oxides nearly always involve the presence of impurity atoms at the grain boundaries it has been shown that the charging can come about intrinsically as well. There has been a report that conducting probe AFM has found a grain boundary in $SrTiO_3$ that is more conducting along its sheet than the bulk crystal.[72] If this is a reproducible result then the possibility that charge inversion has occurred in this boundary should be investigated.

Inversion has been found along a low angle boundary in Ca-doped $YBa_2Cu_3O_{7-x}$, but for mobile charges in the lower Hubbard band.[73] Figure 7.4 shows the schematic result of electron holography measurements which reveal that Ca doping, according to the explanation provided, decreases the width, w, of the normal state insulator + inversion region and thereby accounts for the observed increase in critical current density, i.e. there is more conducting volume about the dislocations in the grain boundary. It should be noted that the measured grain boundary potential is negative, contrary to previous models[74] and expectation of a positive potential.

Ge and Si monocrystals are similar in that they are able to contribute many fewer defects to limit the diffusion length than do grain boundaries. Thus, the diffusion length of these semiconductors can reach levels approaching 1 cm. Most compound semiconductors do not exhibit this quality. The diffusion length of single crystals for these compounds do not exceed about 10 μm. There has been a tendency lately (2004) to ascribe unusual properties to grain boundaries in these compound semiconductors because their grain boundaries effectively have a longer diffusion length than that

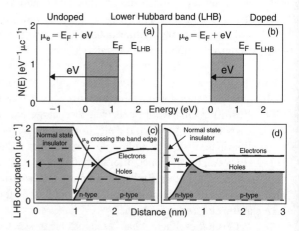

Figure 7.4. Schematic diagram deduced from electron holography measurements showing in a and b the positions of the electron chemical potential in the grain boundary and of the Fermi energy relative to that of the lower Hubbard band for undoped and doped YBCO. Also, c and d show the occupation per unit cell in the lower Hubbard band for electrons and holes for undoped and doped YBCO, respectively. (Modified with permission from M.A. Schofield et al., Phys. Rev. Lett. 92, 195502(2004). Copyright 2004 by the American Physical Society.)

of the bulk crystals. This apparent fact ignores the other fact that their bulk crystals have a smaller diffusion length than Si or Ge. It remains to be shown experimentally whether or not grain boundaries in these compound semiconductors differ from grain boundaries in Si and Ge other than by the ease of forming inversion boundaries.

We will now transfer attention from the consideration of inversion grain boundaries and grain boundary potential barriers to the density and properties of traps and defects that exist along grain boundaries.

Reference to Figure 7.3 and equations 1.5–1.7 reveals that the density of grain boundary trap states affects both the height of the potential barrier and the current that can be transported across the grain boundary. Measurements[4] of these trap densities in poly-Si yield values from 10^{11} to 10^{13} cm^{-2} of grain boundary area. In hydrogenated or deuterated poly-Si these limits drop an order of magnitude (i.e. 10^{10}–10^{12} cm^{-2}). However, this range corresponds to an average for the grain boundaries present in the poly-Si films measured. This is about the same density as found[5] in grain boundaries of multicrystalline silicon, which are known to contain impurities as charge traps. This limit to charge trap densities ($<10^{13}$ cm^{-2}) exists for grain boundaries in diverse materials.[56] The limit may imply that above this limiting trap density the electrostatic energy associated with the introduction of an additional trap becomes larger than the structural free energy decrease associated with the additional trap at the grain boundary since this limit applies to charged defects along grain boundaries also (i.e. the number of atoms per unit area of grain boundary is on the order of 10^{15} cm^{-2}).

The main traps along random grain boundaries in impurity-free poly-Si have been identified by electron-spin-resonance (ESR) to be silicon dangling bonds![1] It is well known that dangling bonds act to trap charge.[2] Here we have an unambiguous characterization of the charge-trapping defects in the poly-Si grain boundaries and these defects cannot be associated with impurities since the ESR signal from a silicon dangling bond is unique to only this defect. Of course when impurities are present their contribution to the trapping of charge cannot be overlooked, i.e. their associated traps overwhelm the intrinsic (non-impurity) trap contribution to the related physical properties. Such a case is found in multicrystalline silicon wafers cut from cast ingots.[3] However, in their absence, dangling bonds are the effective trapping sites along most asymmetric grain boundaries, at least in poly-Si. The dangling bonds in the open grain boundaries of poly-Si are almost equally effective traps to electrons and holes.

However, dangling bonds are not the only trap sites along grain boundaries aside from impurities. EBIC studies as a function of temperature[43] reveal that there are both shallow and deep band gap states present at random grain boundaries in Si that act as recombination centers. The deep gap states originate at dangling bonds but the shallow states are believed to stem from stretched bonds and distorted bond angles. However, it must be recognized that at room temperature recombination at shallow gap states is negligible compared to that at dangling bonds. The population of such states differs between grain boundaries, as one might imagine, being much higher at boundaries that are just a few degrees misoriented from a cusp orientation in a grain boundary energy versus rotation angle plot than at the cusp orientation. Also, $\Sigma 3$ (111) coherent twin and low angle boundaries are populated mainly by shallow-gap states, while high angle boundaries and high Σ boundaries (e.g. $\Sigma 13$, $\Sigma 39$) and random boundaries are populated both by shallow and deep gap states. Although hydrogen passivates many dangling bond and shallow gap states it does not remove all of them by any means.

The results of EBIC and ESR measurements as described above relate to the non-radiative recombination properties of grain boundaries. Another property affected by grain boundaries is the mobility of charge carriers across them. The work of two groups on bicrystals[44] has clarified several aspects of grain boundaries that relate to both recombination and mobility issues. The recombination properties of the tilt and twist grain boundaries support the conclusion described above that different boundaries have different recombination strengths. Further, these boundaries have different quantitative effects on majority carrier mobility. The boundary with the largest deviation from its nominal rotation corresponding to $\Sigma 13$ yielded a much lower traversing conductance (mobility) than one with a small deviation from its nominal rotation angle.[65] This trend correlates to the behavior of grain boundary energy about a cusp position.

Incidentally, the microcrystalline Si produced at the boundary between amorphous Si and microcrystalline Si processing conditions is reported, at least in one paper,[45] to have amorphous regions at the microcrystalline grain boundaries. The ESR spin density of the boundaries in this microcrystalline material, which is produced in an excess of atomic hydrogen so as to passivate the dangling bonds, was $3 \cdot 10^{11} \mathrm{cm}^{-2}$.[45] Hydrogen passivated poly-Si has been reported to have spin densities varying from $2 \cdot 10^{10} \mathrm{cm}^{-2}$ to $9 \cdot 10^{11} \mathrm{cm}^{-2}$.[46] Thus, the special behavior of microcrystalline boundaries in yielding high diffusion lengths and high efficiencies for their grain size is not due to a smaller dangling bond density at the grain boundaries. For this reason, in Chapter I we have chosen the "culprit" responsible for this special behavior to be a diminished capture cross-section typical of a dangling bond in a-Si:H rather than in crystalline Si or its grain boundaries. The electron capture cross-section for a typical poly-Si grain boundary having a grain size of 1 μm is about $3 \cdot 10^{-12} \mathrm{cm}^{-2}$. For a-Si:H it is less than $10^{-13} \mathrm{cm}^{-2}$.

The capture cross-sections of impurity atoms in grain boundaries varies of course, dependent upon the impurity and charge, but the largest hole capture cross-section for a negatively charged impurity is on the order of $10^{-10} \mathrm{cm}^{-2}$.[66]

Grain boundaries also affect significantly other electrical properties of various thin films as described in Chapter I. These effects are related to the characteristics of grain boundaries already discussed, i.e. potential barrier, existence of interface traps for charge carriers, adsorbed impurities. Indeed, grain boundaries are necessary for the efficient operation of certain devices, e.g. varistor.[56]

1.1.3. Transport properties.

Although we have mentioned the enhanced diffusivity along a path within a grain boundary compared to one in the crystal lattice we have not mentioned the phenomenon of diffusion induced grain boundary migration (DIGM). This process may be important in thin films because it is a mechanism for the introduction of foreign atoms into the crystal lattice at temperatures where lattice diffusion is too slow to achieve this effect. A possible example of DIGM occurs during the chlorination of CdTe solar absorber layers. It is reported that Cl in CdTe "recrystallizes" the grains producing a larger grain size. This is an unlikely process. The phenomenon of DIGM (diffusion induced grain boundary migration) is a much more likely process to account for the apparent "recrystallization".[93] In this process the grain boundary diffusant powers the migration of the boundary leaving a residue of itself in the region traversed by the boundary. The observation of Cl within the grains themselves supports this viewpoint. Further, any binding between a Cl atom and a defect at the grain boundary would result in the deposit of such associated complexes into the bulk upon grain boundary migration.

2. Dislocations and stacking faults.

2.1. Electrical properties.

The uncertainty concerning the identity of the charge traps along grain boundaries exists also for dislocations. The reason for this statement is that there is no certainty regarding the absence of sufficient impurities to affect the results in most of the experiments conducted to evaluate the identity of the charge traps. These experiments must be designed to take into account the possible presence of such impurities and only a few of the experiments that have been performed satisfy this criterion.

Another aspect relating to the dislocations must be considered when attempting to evaluate the identity of their charge trapping sites. Namely, whether the dislocations are likely to contain constrictions, kinks, jogs, reconstruction defects, and other possible charge trapping sites.

Werner and Strunk[13] studied the electrical activity of first-order twin boundaries in silicon. They found that although ideally symmetric first-order twin boundaries are not electrically active, boundaries with small deviations from the ideal orientation are electrically active. They found that the sites of this electrical activity are along a$\{112\}/6$ dislocations. These dislocations form a hexagonal network which has sixfold symmetry as observed via TEM and threefold symmetry when observed via EBIC. A geometric analysis shows that this hexagonal network must contain three jogs per cell, which implies that jogs are sites of enhanced

electrical activity. Now strained bonds exist all along the length of these dislocations, yet the electrical activity is found only at the jogs. This observation implies
that strained bonds, per se, are not charge trapping sites. Although the first simple
models of dislocations in silicon predicted the existence of rows of dangling bonds
along dislocation lines, later analysis suggested that such bonds would reconstruct
and disappear. Indeed, the observation noted above that electrical activity is limited
to jogs at the intersection of intersecting dislocations supports the later analysis and
suggests that dangling bonds along dislocations are limited to special sites such as
jogs, i.e. dangling bonds do not exist along straight sections of dislocations.

 For a long time up to the present it has been the belief that the sites acting
to trap charge along dislocations and grain boundaries are extrinsic in nature. One
obvious source for the extrinsic trap sites is related to impurities that segregate along
dislocation lines and along grain boundaries. For a long period in the past impurities
have been ubiquitous in semiconductors. However, it has been possible in recent
times to deposit films free of impurities in which there exists evidence for trapping
and recombination. For example, threading dislocations in one such impurity-free
film of silicon[14] begin to affect the electron mobility at a density of about
$3 \cdot 10^8 \, cm^{-2}$. Such an effect must involve charged dislocations and, hence, charges
trapped at dislocations, and it occurs in a film that has been produced under conditions where no impurities would be expected to be introduced into the film.
Although such traps could be associated with a point defect atmosphere about the
dislocations, an independent study of dislocations in "pure" deformed silicon
reveals that there are trap sites along the dislocations themselves, in addition to
those associated with the point defect atmosphere, identified tentatively as 3-fold
coordinated vacancies along the cores of screw dislocations.[15] Thus, the presence
of traps along dislocations not corresponding to impurities and their precipitates
cannot be ruled out. It should be noted that the dislocations in question are not
in their lowest energy metastable condition of straightness. These dislocations
may well contain jogs along them. Are these jogs to be considered as extrinsic or
intrinsic?

 The implication of this question is serious. The state of the dislocations, in
the absence of impurities, depends upon the prior history of the film. Given the
absence of a high-temperature annealing step in their processing history, the dislocations in most films are likely not to be straight, or jog-free. However, even in the
absence of a high temperature annealing step, misfit type dislocations tend to be
straight, whereas threading dislocations are not straight. Thus, in evaluation of the
results of characterization experiments concerning the nature of traps along dislocation lines it is necessary to know the morphology of the dislocations being examined. In the following we will assign an intrinsic nature to charge trapping sites that
are not impurity related.

 A cathode luminescence (CL) study of scratched and of silicon FZ
crystals that have been deformed along different crystal directions has

provided the conclusions that Lomer–Cottrell dislocations and jogs are the centers of the CL lines that appear at 0.812 and 0.875 eV, whereas the lines at 0.934 and 1.000 eV arise from those parts of the dislocations not involved in slip plane intersections. Further, because of the variation in the CL results from deformation along different crystal axes it was concluded that attribution of the results to impurities is highly questionable.[30] On the other hand, deep level transient spectroscopy studies[31] of dislocations that have moved in CZ and FZ silicon crystals have revealed that the signals obtained depended upon the distance moved by the dislocations and that the increment in the signal strength per given distance moved was higher in the CZ grade crystals (which contain a higher oxygen concentration) as compared to the FZ grade crystals. One consequence of the latter results is that the distance moved by dislocations must be taken into account in interpreting their electrical activity. Thus, we are back to the statement in the first paragraph of this section that there is an uncertainty about the identity of the charge traps along dislocations.

The above fairly describes the situation regarding the question of the electrical activity of dislocations prior to about 1990. Since then new characterization tools and techniques and the development of GaN and its immediate applicability has focussed additional attention on the electrical activity of dislocations. GaN has revealed an anomalous behavior that is worth pursuing here. First, it is well known that GaN LEDs and LDs function with threading dislocation densities exceeding 10^8cm^{-2}.[76] However, these devices involve the luminescence properties of GaN with the significant material property affecting device function involving non-radiative recombination. Indeed, observations[21] reveal little non-radiative recombination in GaN films containing up to 10^{10}cm^{-2} of dislocations. However, there are other reports that reveal non-radiative recombination at threading dislocations.[77] Threading dislocations are not uniquely tilt, or twist, and their character depends upon substrate and processing conditions. Shi et al.[78] using high resolution X-ray diffraction to distinguish between edge and screw dislocations and to measure their densities among the threading dislocations claim to show that both types contributed to non-radiative and radiative recombination but the quantitative contribution depended upon the degree to which the edge tilt angle was dependent upon the screw twist angle and vice versa. However, it is not clear in their work whether the decrease in photo luminescence of the band edge peak that occurs with increase in tilt angle is a consequence of a related increase in twist angle. (Are there samples among those they measured where there exist high tilt angles and low twist angles?) According to other measurements[79] pure edge dislocations are optically inactive. Indeed, it is known and accepted that screw dislocations and mixed dislocations (part screw, part edge) do engage in non-radiative recombination in the absence of impurities.[76,79] It is the action of edge dislocations in this regard that is the controversial issue. Edge dislocations are negatively charged in n-GaN.[80–83] Negatively charged edge dislocations decrease electron mobility when intersecting the electron path but does not affect it when parallel to this path.[20,83] The charge along edge dislocations was reported to be 2 electron/ 0.52 nm.[81] This value needs to be independently verified. The origin of the charge along edge dislocations in GaN is not known. It is known that Ga vacancies cannot account for the charge because they do not exist at equilibrium along edge dislocations.[85] The Ga sites along the dislocation core are occupied by Ga atoms. It has been found that oxygen is ubiquitous in GaN grain boundaries and responsible for the yellow PL peak.[86-88] Does it also populate edge dislocations and charge them negatively?

Screw dislocations are the paths of enhanced leakage currents under reverse bias.[84] In GaN having the wurtzite structure $a/3[11\bar{2}0]$ basal plane dislocations exhibit radiative recombination with a peak at 2.9 eV while screw-type dislocations show non-radiative recombination activity.[79]

CdTe and Cu(In,Ga)Se$_2$ have diffusion lengths which do not exceed about 10 μm. This fact alone suggests that recombination centers are limiting properties in these materials. Further, single crystals do not exhibit larger values of the diffusion length which suggests that grain boundaries are not the site of the limiting recombination centers. One study of a new procedure of growing a single crystal of CdTe found a possible correlation between dislocation density and diffusion length.[89] The reduction of dislocation density increased the diffusion length from 3 to 10 μm. Since the maximum diffusion length in single or polycrystalline films prior to this measurement was 3 μm the implication is that recombination centers at dislocations limit the diffusion length, at least in CdTe. This is the only correlation this writer was able to find between some definable structural characteristic and a measure of recombination activity of an absorber layer of either CdTe or Cu(In,Ga)Se$_2$.

There is sufficient evidence to conclude that Frank partial dislocations bounding oxygen-induced stacking faults (OISF) in silicon are the locii of the electrical activity associated with these stacking faults. There is some question as to whether these partial dislocations in stacking fault tetrahedra produced in regions free from oxygen are electrically active or not. Photoluminescence does not occur at the latter partial dislocations while it has been found to occur at the partial dislocations of OISFs.[16] Further, EBIC observations of OISFs reveal non-radiative recombination at the Frank partial dislocations.[17] One possible interpretation of these observations is that the centers of the radiative recombination processes are associated with the oxygen atoms located at the Frank partial dislocations of the OISFs and that in the absence of the oxygen atoms the Frank partial dislocations do not contain radiative recombination sites. Whether or not the latter contain non-radiative recombination centers is not known from the experiments reported. However, another series of experiments[29] discovered that the photoluminescence intensity from oxygen related sites in semi-insulating poly-Si increased on hydrogenation. The interpretation that the hydrogenation acted to passivate the non-radiative sites along the grain boundaries that competed with the oxygen related radiative sites for charge carriers is consistent with the observations.

The observations reported in Chapter I concerning the effect of OISFs on leakage current can be interpreted to signify that oxygen along the Frank partial dislocations acts to enhance the conductivity in the vicinity of the OISFs probably by enhancing the majority charge concentration locally, i.e. defects associated with the oxygen atoms act as thermal donors. Indeed, the involvement of oxygen in thermal donor activity in silicon has been noted.[18]

Stacking faults, despite the lack of dangling bonds, do affect electrical properties of Si deleteriously, perhaps as a consequence of being a site that segregates impurities.[68] They also act as a sink or trap for point defects and can affect electrical properties via the trapped point defects.[70] Further, the partial dislocations that bound stacking faults may create dangling bonds at grain boundaries that they contact.

Stacking faults have been found to be the sites at which threading dislocations, that are involved in laser degradation, nucleate. The most common origin of stacking faults in semiconductors is a supersaturation of point defects. These defects upon condensation along crystal planes can produce stacking faults. Thus, schemes have been developed to prevent the coming into existence of any such supersaturation during processing of devices.[19]

2.2. Optical properties.

Normally, as noted in Chapter III, dislocations are deleterious for optical properties since they induce non-radiative recombination which acts to reduce radiative recombination. However, dislocations sometimes do harbor radiative recombination centers. These centers, which yield the D series of photoluminescence peaks in silicon, have sparked the hope of achieving light production in silicon, an objective of a project financed by INTAS and involving scientists from Russia, France, United Kingdom, and Italy. At the date of this writing (Nov. 2004) no practical result has been reported although progress toward the objective has been achieved.

Although, it is known that the D series of PL peaks do initiate at or near to dislocations, the microscopic structural environment of the origins of these peaks is still not known. However, the most recent studies conclude that the luminescence does not stem from any bonding structure at dislocations but from another source, which may either be at oxide precipitates near the dislocations,[90] or for Si implanted with B or P, which show enhanced luminescence at room temperature,[91] from some source other than the dislocations themselves. At this writing we are just at the beginning in what will undoubtedly be an intensive effort to develop luminescence in Si. In summary, the luminescence believed to stem from dislocations is an indirect effect, at present from an unknown source at or near the dislocations.

3. Point defects.

3.1. Non-radiative recombination dependent properties.

It should be apparent at this point in the book that properties are affected and even limited by defects. We will first explore a possible answer to the question of what defect is responsible for the limitation to the maximum value of the diffusion length found for the different semiconductor materials. A possible answer is that this maximum diffusion length is defined by the purity level attainable in the semiconductor material. The data in Table 7.1 roughly support this premise. Further support for this premise is provided by the dependence of the charge carrier lifetime and mobility (and diffusion length) on donor and acceptor density in Si down to a density level of $10^{16} cm^{-3}$.[92]

The diffusion length depends not only on the density of the defects providing non-radiative recombination centers but also on their capture cross-sections. Impurity capture cross-sections for majority carrier in single crystal Si vary from about 10^{-16} to $10^{-13} cm^2$. Those for dangling bonds and impurities in grain boundaries in Si are about $3 \cdot 10^{-12} cm^2$ to $10^{-10} cm^2$.

For the polycrystalline elemental semiconductors it is unlikely that the variation in diffusion length in diverse films at a given donor or acceptor level is related to the impurity content since when care is taken the impurity levels can be smaller than the dangling bond densities as determined by ESR that identifies the atom to which the dangling bond is affixed. For example, spin densities ranging from $7 \cdot 10^{16}$ to $10^{18} cm^{-3}$ have been measured in undoped poly-Si. The σN product for dangling bonds in grain boundaries thus exceed about $2 \cdot 10^5/cm$. That for impurities is no larger than $10^{-3}/cm$. However, in large grained multicrystalline Si, produced from rather impure stock, it is possible that the impurity level and dangling bond density are more nearly alike. For example, from Figure 1.8 we note that dislocations begin to affect the diffusion length in large grained

Table 7.1.

Semiconductor	Diffusion length (cm)	Minimum purity* (cm^{-3})
Ge	0.2	10^{10}
Si	0.05–0.1	10^{13}
GaAs	0.001–0.005	10^{16}
CdTe	<0.001	10^{16}
CuInSe$_2$	<0.001	$2.5 \cdot 10^{17}$

* Minimum number of impurity atoms per cm^3 no matter the purification procedure.

multicrystalline Si at a density of 10^4 dislocations per cm². For a maximum of about $5 \cdot 10^6$ active recombination sites per cm of dislocation line this corresponds to a density of $5 \cdot 10^{10}$ recombination active sites per cm³. This value is certainly less than the minimum impurity level in Si, i.e. $10^{13} \, \text{cm}^{-3}$. However, the σN product has a maximum value of about $0.15 \, \text{cm}^{-1}$ while the impurities in the most pure Si have values of the σN product that vary between about 0.001 and $1 \, \text{cm}^{-1}$. The point to be made is that the defect responsible for most of the recombination of charge carriers varies in a given semiconductor material depending upon the processing and other variables. In Si the significant recombination defect is the dangling bond for poly-Si, micro-Si and a-Si, whereas for multi-Si and single crystal Si it is some impurity localized at dislocations or within the crystal lattice in dislocation free material. Note that donors and acceptors can affect the diffusion length in epi-Si devoid of dislocations and grain boundaries. As shown in Figure 7.5, the hole-diffusion length varies with the donor (P) concentration in pure and dislocation free epi-Si. Similarly, the electron diffusion length varies in a like manner in such material with variation in the acceptor concentration.[92a]

For many compound semiconductors it is not possible to give such a clear-cut description of the recombination centers. In most of these materials the impurity level may well be larger than the density of recombination centers. For example, the density of recombination centers in $Cu(In,Ga)Se_2$ varies from 10^{15} to $5 \cdot 10^{15} \, \text{cm}^{-3}$ as the Ga to Ga + In ratio varies from 0.3 to 0.[94] It also increases as this ratio increases beyond the value of 0.3, as shown in Figure 7.6a. The significance of this defect relative to recombination is that its magnitude correlates to the open circuit voltage of solar cells (see Figure 7.6b). The minimum impurity density in this material is $10^{16} \, \text{cm}^{-3}$. Since the impurity content of either In or Ga is on the order of 1 ppma it seems unlikely that the recombination center is due to an impurity. It is more likely related to some defect that is affected by the In or Ga content. This defect density also depends upon processing. The three stage films exhibit lower defect density than the single stage films. The former is represented by the

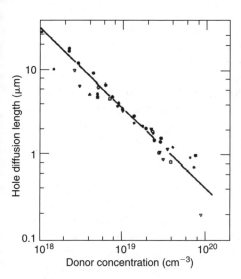

Figure 7.5. Variation of hole diffusion length with donor concentration in epi-Si.
(Reproduced with permission from J.A. del Alamo and R.M. Swanson, Solid State Electron. <u>30</u>, 1127. Copyright 1987 Elsevier.)

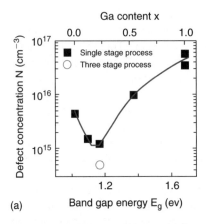

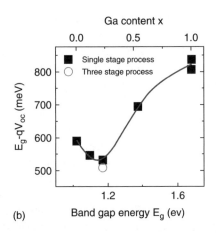

Figure 7.6. (a) Density of recombination defects versus Ga concentration of Ga + In. (Reproduced with permission from G. Hanna et al., Thin Solid Film. 387, p. 71. Copyright 2001 Elsevier.)

Figure 7.6. (b) Open-circuit voltage versus Ga concentration of Ga + In in CuInGaSe$_2$ solar cells. (Reproduced with permission from G. Hanna et al., Thin Solid Film. 387, p. 71. Copyright 2001 Elsevier.)

open circle in Figure 7.6 while the latter produces the filled points. Incidentally, there are other defects in Cu(In,Ga)Se$_2$ that act as recombination centers. However, their population densities are much smaller than the 300 meV admittance spectroscopy step defect that they can be ignored in this discussion.

The literature is very vague concerning the microscopic origin of recombination in CdTe and Cu(In,Ga)Se$_2$. It is possible to find some instrumental observation that correlates to parameters such as short-circuit current and open circuit voltage in CdTe but with no clue as to the microscopic source of the instrumental parameter measured. In this case the latter is the lifetime observed in time-resolved photoluminescence.[97] As already noted for Cu(In,Ga)Se$_2$, where the observable is a step in admittance spectroscopy revealing a defect 300 meV above the valence band that correlates to the loss in open-circuit voltage[94] we still do not have a microscopic description of this defect. Considering the potential leap forward that such knowledge would initiate one might wonder why no strong effort is devoted to its attainment. However, one must admit that this is not an easy problem to solve primarily because it is not possible at this time to be able to vary one structural parameter without varying dozens of other structural parameters in these materials.

A possible path is to introduce specific defects by irradiation and/or implantation. Jasenek and Rau[98a] have studied the effect of electron and proton irradiation on the solar cell parameters of Cu(In,Ga)Se$_2$ and found that the results of electron irradiation could be almost completely explained by an effect on the

open-circuit voltage by means of an increase in the number of recombination center defects, i.e. those defined by the admittance spectroscopy step at 300 meV above
the valence band edge. Characteristics of this defect revealed in recovery experiments involving annealing, light exposure and dark injection of minority carriers
are that the annealing involves an activation *free* energy of 1.05 eV, with a time constant of about 10^{-10} seconds,[98b] light exposure of 100 mW cm^{-2} at room temperature for 3 hours restores more than 90% of pre-radiation efficiency (corresponding
to an appreciable acceleration of the recovery process) and injecting minority carriers by voltage bias in the dark has little effect on recovery.[98c] Apparently, the annihilation of a defect requires the nearby generation of both sign charge carriers. The
defect density increases from a normal level of $4 \cdot 10^{15}$–$8 \cdot 10^{16}$cm^{-3} after a fluence of $5 \cdot 10^{18}$cm^{-2} of 1 MeV electrons. The annealing data suggest that the
recombination defect may involve diffusion of it or its annihilator with an activation
energy of 1.05 eV by about 100 steps or the simultaneous jumping of several atoms
at and about the recombination center in restricted directions for the recombination
defect to annihilate. The observation that light exposure removes the density of
recombination defects to the level existing prior to high energy irradiation and
that injection of minority carriers has no effect implies that both excess minority
and majority carriers are needed for this annealing out process to take place.
Recombination of minority and majority carriers can provide the band gap energy
locally and thereby aid in the activation process needed for the annealing out of the
excess recombination defects.

    Unfortunately, the possibility of determining whether the non-radiative
defect center follows an equilibrium distribution according to an energy of formation or whether it has a non-equilibrium origin, such as an impurity atom or an
extended defect (dislocation), has not as yet been ascertained. The experiment to
accomplish this goal requires determination of the defect density reached upon
long-time annealing (the equilibrium or extrinsic value) at different temperatures
subsequent to irradiation formation of the defects.

    The microscopic identity of the main non-radiative defect center or its
spectrum of centers is important as a potential aid in deducing how to reduce the
associated densities. Such a reduction is one of the two ways of increasing the diffusion length beyond its present limitation of about 3μm. Another potential of
achieving this increase in diffusion length and of the solar cell efficiency is to discover how to reduce the capture cross-section. There have been reports in the literature of Cu(In,Ga)Se$_2$ cells having a barrier to trap filling which reduces the
capture cross-section of the traps to a measly 10^{-23}cm^{-2}.[99] The significance of
this result has yet to be evaluated at the time of this writing. Does this result have
any relation to the properties of the grain boundaries predicted by Persson and
Zunger[54] and/or to the built-in potential[95] in some of these cells?

    We do have some idea of the nature of the significant recombination
defect in GaAs. It is the EL2 defect, which is either an As$_{Ga}$ antisite defect or a

complex involving this defect, that is the dominant non radiative recombination defect in GaAs.[100]

The EL2 defect in GaAs corresponds to a donor level in the middle of the band gap. In semi-insulating material it compensates residual acceptor impurities and accounts for the semi-insulating nature of this material. This defect involves the antisite defect As_{Ga}, which has two extra electrons not used for the covalent bond and thus is a double donor. The first donor energy level for the EL2 defect is the same as for the As_{Ga} antisite defect ($E_V + 0.52$ eV), while the second energy level ($E_V + 0.75$ eV) differs from that for the antisite defect ($E_V + 0.70$ eV) only slightly. It is believed that the EL2 defect is a complex of the As_{Ga} and an arsenic interstitial that is bound to the antisite defect. It is known that the EL2 defect can be transformed into a metastable defect by optical excitation at low temperatures <130 K in what is known as the persistent photoquenching effect. The EL2 trap level disappears completely from the band gap upon this excitation, but can be reactivated by annealing above 130 K. In the metastable state a monovacancy is associated with the defect.[101]

Among the properties that the EL2 defect confers in semi-insulating GaAs are a short carrier lifetime, high carrier mobility, and high dielectric breakdown strength. Layers containing these defects are used as buffer layers for MESFET structures and ultrafast photo-detectors.

The antisite defect may not be as benign as implied above. In Chapter III we noted that in the operation of various LEDs that product point defects are generated by non-radiative recombination at parent point defects, that the product point defects can act as non-radiative recombination centers, where such recombination leads to the generation of additional point defects and a continuation of the sequence. What is significant in this story is that the energy released by non-radiative recombination locally at a point defect can generate additional point defects that have the same property of reproducing themselves upon non-radiative recombination in their immediate vicinity.[102] A clue to the identity of these "cloning" type of defects is that interstitial-type Frank microloops have been observed after operation of the LED, but not before such operation. The latter observation suggests that interstitials are involved in the formation of a product defect from a parent defect.

The possibility that the defect responsible for the rapid degradation of laser diodes in the III–V system may act to regenerate itself has been suggested by Dow and Allen.[33] In their model non-radiative recombination at a dangling bond induces the formation of another dangling bond. They are deliberately vague as to how this occurs in detail. Let us examine possible mechanisms for the production of interstitials that can condense to produce interstitial-type Frank microloops. A simple mechanism stemming from Dow and Allen's suggestion is the formation of a Frenkel defect in the vicinity of a vacancy. Now, whether a Ga or an As interstitial is produced by expelling the atom from its proper site, at least four Ga and four As dangling bonds will be generated. For this process to be possible the energy available from the recombination process must exceed the sum of these dangling bond energies, i.e. the average dangling bond energy must be less than one-eighth the band-gap energy. The energy of an unrelaxed dangling bond energy in GaAs may be estimated from the cohesive energy to be approximately 0.93 eV. Thus, the energy to form an unrelaxed Frenkel defect in GaAs should equal about 7.4 eV. Relaxation processes should reduce this formation energy significantly, but, certainly not down to the level of 1.45 eV, the energy available from the recombination process.*

On the other hand, production of a Frenkel defect involving a Ga vacancy and a Ga interstitial in a two step process that involves two successive recombination events may be possible. Consider that an As_{Ga} antisite defect expels an As atom into an interstitial position upon non-radiative recombination at the As_{Ga} antisite defect. The bond energies associated with the antisite defect total about 1.5 eV less than the total bond energies normally at this site. Thus, the possibility for recombination energy promotion of the antisite As to an interstitial position becomes more probable. Also, subsequently, another non-radiative recombination event at the As interstitial can act to exchange it with a Ga atom at a Ga site to form a Ga interstitial and an As_{Ga} antisite defect. Thus, although the result is the same as for the Dow–Allen mechanism, two recombination events provide the required energy instead of just one. In this way the original defect is regenerated while a new interstitial is also generated by the action of two recombination events. Thus, there need be no saturation of the degradation process, which involves the net production of Ga interstitials and, upon condensation of the latter, interstitial-type Frank dislocation loops. If the energetics involved for the various defects were known it would be possible to predict whether the degradation sequence described could be energized by the recombination process. Roughly, this concept predicts that the smaller the energy gap of the semiconductor, the less energy is available to drive the defect reactions required for the laser degradation. This prediction is consistent with the observations that the degradation does not occur in InGaAsP but does occur in GaAlAs.

An experimental study[103] of recombination enhanced diffusion in GaAs found that recombination at the recombination center emits a point defect that enhances the diffusion of Be solute atoms and also annihilates the recombination center. A candidate proposed for this point defect is the Ga interstitial. The above mechanism may be consistent with the experimental data providing that some of the As interstitials produced by the recombination induced activation diffuse to vacancies on As sites. This process acts to reduce the number of EL2 defects and thereby satisfy the experimental observation that the process results in annihilation of recombination centers.

---

* For Si with a dangling bond energy of 1.17 eV deduced from the cohesive energy, the lowest calculated value of the energy of a Frenkel defect is 6 eV, or 64% of the thermodynamically deduced value, whereas the lowest experimental value is 55% of the latter value.[40]

A case where a defect is well defined and produces an effect on a property that anneals out with time at room temperature is that represented by the DX defect in AlGaAs and other semiconductor compounds. The property is that of persistent photo conductivity (PPC), which appears as a photo-induced electron density that does not anneal out when the sample is rapidly cooled to below room temperature in the dark and held there. This effect is illustrated in Figure 7.7. As shown, the photo-induced electron density can be quenched-in by rapid cooling to below a critical temperature. Microscopically,

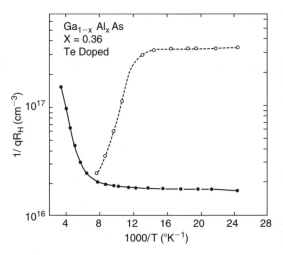

Figure 7.7. Reciprocal Hall coefficient as a function of reciprocal temperature. Data taken for sample in the dark for solid curve; the dashed line corresponds to data taken after the electron density has been saturated by photo-excitation at low temperature. (Reprinted with permission from Nelson, Appl. Phys. Lett. <u>31</u>, 351(1977). Copyright 1977 American Institute of Physics.)

the entity that causes the effect is a donor. The change that is photo-induced is a change of state of this entity rather than the creation of a defect. This concept is illustrated in Figure 7.8. As shown, there are two states of the atoms about the donor atom. In the normal donor state the donor atom occupies a Ga site and is coordinated to four As atoms. In the DX state the donor atom is coordinated to three As atoms with the donor atom displaced to an interstitial-like site and a Ga vacancy at the site the donor atom previously occupied. The DX state is at the lowest energy, it is a deep-level state. With the donor atom in the d^0 state it is a shallow level donor. Transitions between these two states can be produced by pressure and composition. Photoionization of a DX center leads to electrons occupying the potential well with the minimum at 0 in the defect configuration coordinate diagram illustrated in Figure 7.9. The non-ionized DX center has its potential well at the position Q_R in the figure. The ionization energy is denoted by E_n. For the electrons to be captured by the ionized DX center and for it to return to the non-ionized configuration an energy barrier equal to E_B must be overcome. This energy barrier is responsible for the PPC effect in AlGaAs and other GaAs based alloy semiconductors. In the ionized condition the defect is metastable.

There are other semiconductor compounds that exhibit PPC. In particular, light with a photon energy of 1.76 eV can excite a metastable increase in the dark conductivity of CdS at 77°K.[27] Also, many other II–VI alloys exhibit the PPC effect. However, it is not certain that the DX type defect is responsible for the effect in these materials.

An interesting property resulting from manipulation of the DX centers in GaAlAs and CdZnTe has been reported.[28] The excitation of the DX centers to shallow states results in a change in the index of refraction in the vicinity of these excited centers. Further, the PPC associated with these centers can be localized to regions as narrow as 0.1 μm. The electrons do not wander far from the ionized DX centers. Hence, optical diffraction gratings can be encoded in these materials at low temperatures for holographic recording. They have better photo refractive properties than $LiNbO_3$. Unfortunately, these properties cannot be maintained at room temperature. What is needed is a DX center with a higher barrier between the excited state and stable state. It is possible that the ionic compound CdF_2 is the desired host for this defect for this application.[104] Another interesting, but potential, application of bistable defects, such as the DX

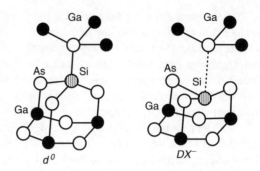

Figure 7.8. Stable configurations for a Si-donor impurity in GaAs. (Modified with permission from Chadi and Chang, Phys. Rev. B39, 10063(1989). Copyright 1989 by the American Physical Society.)

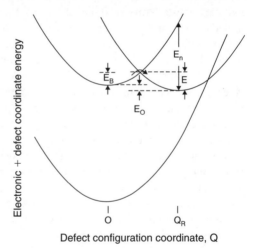

Figure 7.9. Proposed configuration-coordinate diagram for DX centers. (Modified with permission from D.V. Lang and R. Logan, Phys. Rev. Lett. 39, 635(1977). Copyright 1977 by the American Physical Society.)

center, is in memory storage, providing that such a center could be switched by an externally applied signal and that its configuration can be read. The bistable character of the DX defect has inspired others to search for similar bistable defect behavior with results that suggest the existence of many similar defects in a variety of other semiconductor thin films.

These are not the only projects for using defects to engineer properties. Another is in the field of shallow junctions. In the desire to prevent the diffusion of boron at junctions use is made of self-implantation of Si to produce vacancies that will form complexes with boron and anchor them so that they do not indulge in diffusive processes. The reverse process has been observed where Be doping in GaAs acts to maintain the concentration of As_{Ga} defects independent of annealing temperature, presumably by forming a complex that requires too high a temperature to dissociate.

The dangling bond is a significant non-radiative recombination center in a number of hosts. We will first discuss the dangling bond in hydrogenated amorphous silicon. In the first book to this series we concluded that reduction of dangling bond density in a-Si:H requires reduction in the energy of the weak bonds and their population density. The reason a reduction in dangling bond density is desired is that usually, but not always, properties improve with such a reduction. We have already in earlier chapters cited the evidence relating dangling bond density to properties. We have found that the role of the dangling bond as the predominant non-radiative recombination center is the basis of its effects on a variety of electronic properties. Since it is desirable to reduce the dangling bond density and since this density depends upon the maximum energy and densities of weak bonds it is a reasonable decision to pay attention to the weak bond spectrum here.

Reitano et al.[105] found that the FWHM of the Raman $480\,cm^{-1}$ peak which is a measure of the bond angle distortion had a value of 68 in annealed a-Si. The value found[106] in device quality a-Si:H produced with hydrogen dilution is about the same at 66. This is an unexpected result. The conclusion following from it is that the bond angle deviation from the normal sp^3 configuration is about the same in the hydrogenated film as in the unhydrogenated film. Yet the dangling bond density and weak bond state density are much larger in the unhydrogenated film. Thus, hydrogen reduces the weak bond density but does not affect the bond angle deviation from the sp^3 configuration. This result suggests to me that the reaction involved between H and the weak bonds does not have a bias to react with the highest energy bonds preferentially. This is a reaction that occurs because of happenstance, such as the propinquity of reactants, rather than a thermodynamic gradient driven one. The only form of hydrogen in the a-Si:H that is positioned to be involved in such a reaction is molecular hydrogen situated in empty interstitial tetrahedral sites.[107]

Before we consider how hydrogen acts to decrease the weak bond density let us investigate certain aspects of the deposition of these films. We have no good idea as to how during deposition of a-Si:H a surface equivalent of about 10^{20} dangling bonds per cm^3 is diminished to much less than $10^{16}\,cm^{-3}$ in the bulk. Incidentally, this surface equivalent dangling bond density is the same for both a-Si and a-Si:H. In amorphous silicon this equivalent density ends up as the actual bulk dangling bond density, whereas in hydrogenated amorphous silicon it is appreciably diminished. Obviously, rearrangements occur below the surface that annihilate dangling bonds and, concomitantly, weak bond density. Molecular dynamic simulations of the growth process reveal no subsurface activity when the precursor is the Si–H$_3$ radical.[109] Only when the precursor is Si–H is there much reaction with subsurface atoms and then there is no indication of rearrangements of bonded configurations.[110] However, the size of the cell in MD hinders observation of dangling bonds which exist[116] at a surface concentration level of 10^{-2} while the cell consists of not much more than 200 atoms for a concentration minimum of $5 \cdot 10^{-3}$. Thus, the most likely mechanism of achieving the diminution of surface equivalent dangling bond density to that observed involves H reaction with the excess dangling bonds although a fraction of the excess dangling bonds that are nearest-neighbors could form Si–Si bonds as they are submerged below the next layer (but they do not do this in a-Si). We do not know whether the H that accomplishes this initial reaction during growth is interstitial H$_2$ or interstitial H. Nor do we know whether H diffusion is necessary for the reaction with the dangling bonds or is propinquity of adjacent hydrogen also involved in this reaction. We know that both the metastable equilibrium values of dangling bond density and weak bond density (Urbach slope) depend upon the hydrogen concentration in the film. The simultaneous dependence of dangling bond density and weak bond density on the hydrogen concentration is revealed in Figures 7.10 and 7.11. (The effect of anneal temperature above 400°C is to decrease the hydrogen concentration in the film.)

How does H reduce the weak bond density without increasing the dangling bond concentration? One way is to replace weak bonds with Si–H bonds at

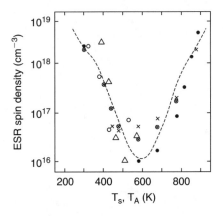

Figure 7.10. ESR spin density (dangling-bond density) versus substrate temperature T_S or annealing temperature T_A in undoped α-Si:H. (From M. Stutzmann, Phil. Mag. B60, 531(1989) with permission. Copyright 1989 Taylor & Francis Ltd.)

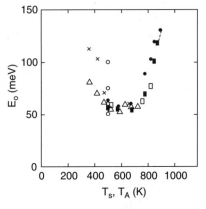

Figure 7.11. Data plots of the Urbach tail slope, E_o, versus the substrate or annealing temperatures, T_S, T_A. (From M. Stutzmann, Phil. Mag. B60, 531(1989) with permission. Copyright 1989 Taylor & Francis Ltd.)

the weak bond sites. The sum of the weak bond population density and that of the Si–H bond population in device grade a-Si:H is $5 \cdot 10^{19} + 4 \cdot 10^{21}\,\text{cm}^{-3}$ whereas that of the weak bond population density in a-Si is about $2 \cdot 10^{20}\,\text{cm}^{-3}$. This confrontation shows that the concept that Si–H bonds replace weak bonds is not inconsistent. However, it suggests that only a fraction of the Si–H bonds are replacements of weak bonds. In particular, according to the math $3 \cdot 10^{20}\,\text{cm}^{-3}$ Si–H bonds have replaced weak bonds, i.e. at the most two Si–H bonds replace one weak bond. Thus, slightly less than 10% of the Si–H bonds in device grade a-Si:H are used to replace weak bonds. Further, this excess in Si–H bonds is more than sufficient to correspond to the equivalent volume excess of surface Si dangling bonds if all of them are submerged without bonding to other Si atoms.

The correlation found[111] between the amount of molecular hydrogen existing in T type interstitial sites and the photo response product shown in Figure 7.12 suggests that molecular hydrogen may be the agent that controls the weak bond population density. Consider, for a given trap and charge carrier the product mobility · lifetime · dangling bond density is a constant. (This relation actually follows from the relation between these quantities and the capture cross-section of the dangling bond trap.) As shown in Figure 1.7 the capture cross-section is roughly constant for a group of samples varying, however, between groups. The

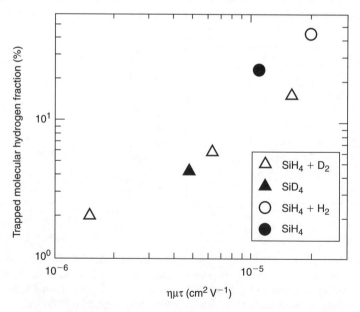

Figure 7.12. H_2 in T sites versus photo response product. (From D.J. Leopold et al., MRS Symp. Proc. 715, A1.1.1(2002) with permission.)

defining identity of the group is the combination of trap and charge carrier being trapped. If we assume that the capture cross-section is constant for the data in Figure 7.12 then these data suggest that the photo response product varies inversely as the dangling bond density and consequently, that the increase in molecular hydrogen concentration in interstitial T sites has decreased the dangling bond density, presumably by decreasing the weak bond population. This result is consistent with the reaction

$$H_2 + Si^\wedge Si = 2\ Si\text{–}H$$

and its mass action relation

$$[H_2][Si^\wedge Si] = [Si\text{–}H]^2\ K$$

and the conservation of hydrogen. For a given total hydrogen content in a film the only way this relation is satisfied when $[H_2]$ increases is for $[Si^\wedge Si]$, the concentration of weak bonds, to decrease. How is it possible to vary $[H_2]$ in the interstitial T sites?

The concentration of molecular hydrogen in T sites may be controlled by an equilibrium between the partial pressure of H_2 in the growth environment adjacent to the film surface and H_2 in T sites. Hence, to increase $[H_2]_T$ the partial molar heat of solution of $[H_2]_T$ must become a more negative quantity. In structural terms this means making molecular hydrogen in T sites a more stable configuration. One possible way to accomplish this objective is to increase the specific volume of the a-Si:H network. (This assumes no bonding between the molecular hydrogen and the surrounding environment.) This feature will reduce the strain energy in the Si network associated with the expansion occasioned by the introduction of a hydrogen molecule into a T type interstitial site. Perhaps the introduction of inert atoms such as He will expand the network and increase the solubility of molecular hydrogen in T like sites. It is known that He dilution of silane acts to lower the bonded H content as measured from IR absorption spectra.[108] Also, it has been found that He acts to decrease the band gap, to decrease the specific volume, to increase the photo response product, and to decrease the weak bond population.[112] Nothing is known about the effect of He dilution of silane on the molecular hydrogen concentration in T like sites. The reduction of specific volume is believed to be a consequence of the reduction in the bonded H content. However, a-Si:H is known to be a heterogeneous material. Perhaps, in the vicinity of the He there is an expansion of the network while the removal of clustered Si–H bonds in vacancies and voids by the removal of the latter acts to densify the network elsewhere with a net increase in the overall density.

As is apparent from the above we have entered the realm of speculation. There is much to be learned. Perhaps the speculation will fuel research to remove the areas of ignorance.

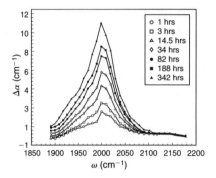

Figure 7.13. Increase in absorption of Si–H stretching IR peak with exposure to white light of 0.4 W/cm^2.a$_o$ = 800 cm^{-1}. (Modified with permission from Y. Zhao et al., Phys. Rev. Lett. 74, 558(1995). Copyright 1995 by the American Physical Society.)

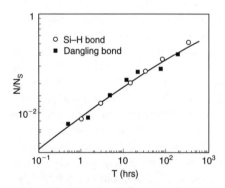

Figure 7.14. Normalized dangling bond density and Si–H concentration versus time of exposure to white light. (Modified with permission from Y. Zhao et al., Phys. Rev. Lett. 74, 558(1995). Copyright 1995 by the American Physical Society.)

Incidentally, there is some evidence which is consistent with the concept of molecular hydrogen in a-Si:H. Using a sensitive technique Yiping et al.[113] have found that white light exposure increases the 2000 cm^{-1} IR Si–H stretching mode absorption peak as shown in Figure 7.13. This result has been interpreted as an effect of light exposure in increasing the concentration of Si–H bonds at a rate proportional to the increase in dangling bonds. This increase cannot be a result of an instantaneous equilibrium between dangling bonds and Si–H bonds since the annealing out of the Si–H bonds occurs over a time period with an activation energy of about 0.9 eV. However, the generation does not involve a time dependence associated with H diffusion. The generation occurs at the same rate as the generation of dangling bonds (see Figure 7.14). If the increase in the intensity of the 2000 cm^{-1} peak really represents an increase in Si–H concentration then this must be a light exposure induced recombination enhanced effect of a reaction that produces such bonds. We have already discussed this reaction above. To my knowledge it is the only reaction that can satisfy the instantaneity of the reaction, i.e. H$_2$ next to a weak bond can react with it without requiring diffusion of H or for that matter that can provide the H needed to produce *additional* Si–H bonds.

There is still another experimental result[117] that is consistent with the concept of molecular hydrogen presence in a-Si:H and which verifies the formation of Si–H bonds on exposure to light, with the additional information that these bonds are paired and separated by 2.3 Å. The use of ^{1}H NMR revealed this pair of Si–H bonds formed upon light exposure of a-Si:H. The authors claimed that this observation could not be due to the H atoms in a H_2 molecule. However, they observed Si–H bonds which could be formed from the reaction between a H_2 molecule and two weak bonds stemming from a Si atom that is a nearest-neighbor to the H_2 molecule in a T site. The two Si–H bonds would then be separated by the distance between neighboring Si atoms or 2.3 Å, the observed distance!

To return to the theme of the effect of structure on properties we should note that the state of metastable equilibrium requires a proportionality between the dangling bond and the weak bond concentrations. However, it has long been known that the dangling bond population can vary at a constant Urbach slope or weak bond population. For example, see the scatter in dangling bond density at constant Urbach slope in Figure 7.15. (This scatter in dangling bond values is more than the ESR measurement error.) Another example is a variation in the photo response product at constant Urbach slope with solar cell properties varying monotonously with the photo response product. Here we enter into a regime I have avoided until now which is that of the interaction between charges and defects. So I will avoid it.

Is it possible for weak bonds to react with other solute than H? Of course they can. It appears that B acts to affect the weak bond density in that increasing B content in a CVD a-Si:H deposited film increases the Urbach inverse slope energy as shown in Figure 7.16. B acts to decrease the specific volume of the amorphous network and it is hypothesized that it acts to increase the bond distortion and bond angle strain. But B is a much larger atom than H and it bonds

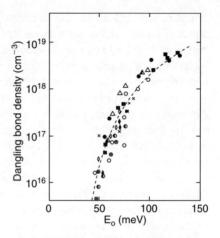

Figure 7.15. Dangling bond density as a function of the Urbach slope. (Reproduced with permission from M. Stutzmann, Phil. Mag. B60, 531(1989). Copyright 1989 Taylor & Francis Ltd.)

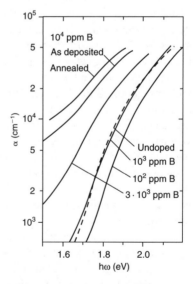

Figure 7.16. Room temperature absorption spectra of boron doped glow discharge a-Si:H. (Reproduced with permission from J. Ristein and G. Weiser, Solar Energy Mater. 12, 221. Copyright 1985 Elsevier.)

simultaneously at least to three Si atoms, whereas H bonds only to one. The hopeful aspect of Figure 7.16 is that it shows that B affects the weak bond density, albeit for the worse. Perhaps Li as a single valent solute may act similarly to H to remove weak bonds forming SiLi bonds or act together with H to reduce the weak bond population? I am not aware of whether it has been tried or not.

Another possible concept is to replace distorted and bent bonds with non-distorted and non-bent bonds by replacing the Si atoms at these bonds with other quadrivalent atoms that convert the strained bonds to unstrained ones, i.e. try a mix of quadrivalent atoms with different distortion free bond lengths at a concentration level about that of the weak bond concentration.

It is believed by this writer that the current technique of using hydrogen dilution to produce a-Si:H at the protocrystalline stage accomplishes the result of reducing the weak bond maximum bond angle distortion by nucleating crystalline nanograins at the sites of the most energetic weak bonds. The hope in this process is not to generate additional and more distorted weak bonds at the interfaces between these nanocrystals and the surrounding amorphous matrix. An argument in support of this interpretation is based on the result for the Raman $480\,cm^{-1}$ TO peak FWHM for the a-Si network dependence on H dilution in the range where amorphous films are produced, as shown in Figure 7.17, i.e. as the crystalline content in the film increases from zero, the a-Si TO peak FWHM decreases from about $80\,cm^{-1}$ to about $67\,cm^{-1}$. Stated plainly, the formation of crystals in the a-Si:H network decreases the maximum bond angle distortion from about $10°$ to about $8°$.[115] Is there a correlation between this decrease in maximum bond angle distortion and the diminution of the Staebler–Wronski instability in the a-Si:H films now produced with H dilution? I also wonder whether application of techniques from the field of nucleation to maximize nucleation density and minimize nuclei growth may extend the benefits of H dilution processing to yield still more improved properties in a-Si:H.

So much for the weak bond density and bond angle maximum distortion. Now we might ask is there anything that can be done to modify the ability of a dangling bond to promote non-radiative recombination. Apparently, the dangling bond acts as a trap for a charge, probably an electron.[118b] The obvious solution is to

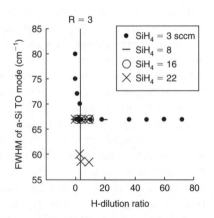

Figure 7.17. The decrease in FWHM for the a-Si peak for the filled circle group all occurred as the amount of crystalline Si increased from 0. (Reprinted with permission from J. Appl. Phys. <u>94</u>, 2930(2003). Copyright 2003 American Institute of Physics.)

keep electrons and holes apart from each other. This is the function of a potential gradient which is accomplished at inversion grain boundaries that set up separate parallel regions in which electrons can migrate separately from holes. Also, this is a deliberate strategy in nipi devices where the modulation of dopants is used to minimize recombination by setting up separate layers of electrons and holes.[114] What is needed is a processing technique that makes these layers align perpendicular to the plane of the film, i.e. a spinodal decomposition with modulation in the plane of the film or a pattern produced from a reaction-diffusion situation as described in Chapter VIII of the revised edition of Volume I in this series or the use of a scheme pioneered by Greenham et al.[120] to provide intermixed electron affinity and hole affinity phases in a composite layer.

I have used up much space in this chapter in a discussion of how to control the dominant defect density or its effects on properties. This has necessarily led to a discussion of the subject of processing and the theme of this book is the effects of structure on *properties* not processing. For this I apologize, but note that the effects of the structure on properties were considered in detail in prior chapters. This chapter has focussed on the prominent defects that control properties. For the case of the dangling bond in a-Si:H, although implied but not stated previously in an explicit manner, the dominant defect is the weak bond.

In hosts other than a-Si and a-Si:H the dangling bonds are not generated at weak bonds. In poly-Si grain boundaries dangling bonds are generated by geometric requirements, i.e. at screw dislocation jogs and the like. Indeed, the characteristics of dangling bonds in poly- and micro-Si grain boundaries differ from those in a-Si or a-Si:H.[118] The capture cross-section of dangling bonds in a-Si:H differs from that for dangling bonds in grain boundaries, as already noted. In a-Si:H nearby weak bonds first capture an electron which then subsequently is captured by a nearby dangling bond.[118b] In the grain boundaries of crystalline Si it appears that the dangling bond also captures electrons directly without the intervention of nearby weak bonds.[118b] Unless the texture of films can be controlled precisely so as to eliminate any deviation of a grain boundary from a precise tilt orientation dangling bonds will exist along grain boundaries. Further, the intersection of partial dislocations bounding stacking faults with grain boundaries will introduce dangling bonds at the latter even if they had the ideal tilt orientation. (Avoidance of {111} growth surfaces minimizes the generation of stacking faults. However, this act implies the use of conditions that increase the cost of the films.) Hence, the pragmatic strategy to deal with this situation is to separate photo generated electrons and holes into separate regions of space where these charges can migrate without recombination. This possibility has already been discussed in this chapter.

Another type of point defect which we have not yet discussed in this series is the valence alternation defect. This defect comes in pairs, the lowest energy configuration for these defects. The pair consists of an over-coordinated atom and an under-coordinated atom conserving the number of covalent bonds.

These valence alternation pairs (VAP) are to be found in chalcogenide glasses and exist at a density of about $5 \cdot 10^{15}$–10^{17} cm^{-3}.[119] Reference 119 provides a detailed and complete summary of the characteristics of these defects and their effects on properties. The localized nature of the reaction involved in producing and annihilating these defects resonates in my mind with the localized nature of the reaction between molecular hydrogen and a weak bond to produce a pair of Si–H bonds 2.3 Å apart. However, in the latter there is no change in coordination of the atoms while such a change is characteristic of the former.

References

1. N.M. Johnson, D.K. Biegelsen and M.D. Moyer, Appl. Phys. Lett. 40, 882(1982). (The assignment of the g = 2.005 signal to grain boundary dangling bonds has been questioned by M.M. de Lima Jr. et al., Phys. Rev. B65, 235324(2002). However, in view of its amplitude varying with grain boundary area, that this signal is not present in crystal Si and that it has the same g value as the dangling bond in a-Si the position of de Lima et al. as stated by them seems weak to this bystander.)
2. R.A. Street, **Hydrogenated Amorphous Silicon**, Cambridge University Press, 1991.{Aq}
3. I. Perichaud and S. Martinuzzi, MRS Symp. Proc. 262, 481(1992).
4. J.E. Werner and M. Peisl, MRS Symp. Proc. 46, 575(1985); R. Podbielski and H.J. Möller, in **Poly-Micro-Crystalline and Amorphous Semiconductors**, eds. P. Pinard and S. Kalbitzer, Les Editions de Physique, Les, Ulis, 1984, p. 365; R. Podbielski and H.J. Möller, in **Proceedings of the 13th International Conference on Defects in Semiconductors**, eds. L.C. Kimerling and J.M. Parsey, TMS, AIME, Warrendale, PA, 1985, p. 435.
5. M. Aucouturier, A. Broniatowski, A. Chari and J.L. Maurice, in **Polycrystalline Semiconductors**, eds. H.J. Möller, H.P. Strunk and J.H. Werner, Springer Proceedings in Physics, Vol. 35, 1989, 64.
6. Y. Hayamizu, S. Ushio and T. Tanenaka, MRS Symp. Proc. 262, 1005(1992).
7. H.J. Möller, Solar Cells 31, 77(1991).
8. L.L. Kazmerski, in **Polycrystalline Semiconductors**, eds. H.J. Möller, H.P. Strunk and J.H. Werner, Springer Proceedings in Physics, Vol. 35, 1989, 96.
9. R.D. Tomlinson, A.E. Hill, M. Imanieh, R.D. Pilkington, A. Roodbarmohammadi, M.A. Slifkin and M.V. Yakushev, J. Electron. Mater. 20, 659(1991).
10. N.H. Nickel, N.M. Johnson and C.G. Van de Walle, Phys. Rev. Lett. 72, 3393(1994).
11. Y.V. Gorelkinskii and N.N. Nevinnyi, Sov. Tech. Phys. Lett. 13, 45(1987); B. Holm, K. Bonde Nielsen and B.B. Nielsen, Phys. Rev. Lett. 66, 2360(1991).
12. K. Matsuda-Jindo, **Polycrystalline Semiconductors**, eds. H.J. Möller, H.P. Strunk and J.H. Werner, Springer Proceedings in Physics, Vol. 35, 1989, p. 52.
13. J. Werner and H. Strunk, J. de Phys. (Paris) 43, 89(1982).
14. K. Ismail, J. Vac. Sci. Tech. b14, 2776(1996).
15. D. Cavalcoli, A. Cavallini and E. Gombin, Phys. Rev. B56, 10208(1997).
16. Y.H. Qian, J.H. Evans, L.F. Giles, A. Nejim and P.L.F. Hemment, MRS Symp. Proc. 378, 629(1995).

17. A. Ourmazd, P.R. Wilshaw and G.R. Booker, Physica B116, 600(1983).
18. C.S. Fuller, J.A. Ditzenberger, N.B. Hannay and E. Buehler, Phys. Rev. 96, 883(1954); W. Kaiser, H.L. Frisch and H. Reiss, Phys. Rev. 112, 1546(1958).
19. S.-Y. Shieh and J.W. Evans, MRS Symp. Proc. 262, 981(1992).
20. T.-B. Ng, J. Han, R.M. Biefeld, J.C. Zolper, M.H. Crawford and D.M. Follstaedt, MRS Symp. Proc. 482, (1998).
21. F.A. Ponce, B.S. Krusor, J.S. Major, Jr., W.E. Plano and D.F. Welch, Appl. Phys. Lett. 67, 410(1995).
22. R.J. Nelson, Appl. Phys. Lett. 31, 351(1977).
23. D.J. Chadi and K. Chang, Phys. Rev. B39, 10063(1989).
24. C.H. Henry and D.V. Lang, Phys. Rev. B15, 989(1977).
25. D.V. Lang and R. Logan, Phys. Rev. Lett. 39, 635(1977).
26. D.V. Lang, R. Logan and M. Jaros, Phys. Rev. B19, 1015(1979).
27. A.L. Fahrenbruch and R.H. Bube, J. Appl. Phys. 45, 1264(1974).
28. D.J. Chadi and C.H. Park, Mater. Sci. Forum, 196–201, 285(1995).
29. S.S. Ostapenko, A.U. Savchuk, G. Nowak, J. Lagowski and L. Jastrzebski, Mater. Sci. Forum, 196–201, 1897(1995).
30. T. Sekiguchi and K. Sumino, Ibid., p. 1201.
31. O. Kononchuk, V. Orlov, O. Feklisova, E. Yakimov and N. Yarykin, Ibid., p. 1183.
32. See, for example, H.J. Möller, **Semiconductors for Solar Cells**, Artech House, 1993, p. 86.{AQ}
33. J.M. Langer and W. Walukiewicz, Mater. Sci. Forum, 196–201, 1389(1995); D.D. Nolte, Solid State Electron. 33, 295(1990); J.D. Dow and R.E. Allen, Appl. Phys. Lett. 41, 672(1982).
34. J. Tersoff, Phys. Rev. Lett. 52, 465(1984); Phys. Rev. B30, 4874(1984).
35. M. Lannoo and P. Friedel, **Atomic and Electronic Structure of Surfaces**, Springer-Verlag, 1991.{AQ}
36. D.L. Staebler and C.R. Wronski, Appl. Phys. Lett. 31, 292(1977); J. Appl. Phys. 51, 3262(1980).
37. A. Skumanich, N. Amer and W. Jackson, Phys. Rev. B31, 2263(1985); M. Stutzman, Phil. Mag. B60, 531(1989).
38. J.-H. Yoon, J. Appl. Phys. 74, 1838(1993).
39. N.H. Nickel, MRS Symp. Proc. 378, 381(1995).
40. T. Sinno and R.A. Brown, MRS Symp. Proc. 378, 95(1995); H. Bracht, N.A. Stolwijk and H. Mehrer, in **Semiconductor Silicon**, eds. H.R. Huff, W. Bergholz and K. Sumino, **Electrochemical Society**, Pennington, NJ, 1994.
41. K.J. Chang and D.J. Chadi, Phys. Rev. Lett. 62, 937(1989).
42. Dopants behave in the manner discussed for shallow charge sources to segregate to grain boundaries and to provide a charge barrier to minority charge carriers attempting to cross the grain boundaries.
43. S. Tsurekawa and T. Watanabe, Solid State Phenom. 93, 333(2003).
44. A. Fedotov et al., Solid State Phenom. 67–68, 15(1999); J.H. Werner, Inst. Phys. Conf. 104, 63(1989).
45. M.M. de Lima, Jr. et al., Phys. Rev. 65, 235324(2002).
46. N.H. Nickel, G.B. Anderson and R.I. Johnson, Phys. Rev. B56, 12065(1997); N.M. Johnson, D.K. Biegelsen and M.D. Moyer, Appl. Phys. Lett. 40, 882(1982).

47. M. Koyama, Model. Simul. Mater. Sci. Eng. 10, R31(2002); O.B. Hardouin Duparc and M. Torrent, Mater. Sci. Forum 207, 221(1996); M.F. Chisholm et al., Solid State Phenom., 67, 3(1999).
48. P.V. Evans, D.A. Smith and C.V. Thompson, Appl. Phys. Lett. 60, 439(1992).
49. Y. Aya et al., Bull.Am. Phys. Soc. 47, No.1, Part 2, 873(2002).
50. Y. Aya et al., Solid State Phenom., 93, 339(2003).
51. M.F. Chisholm et al., Phys. Rev. Lett. 81, 132(1998).
52. F. Cleri et al., Phys. Rev. B57, 1708(1998).
53. N.H. Cho and C.B. Carter, J. Mater. Sci. 36, 4511(2001).
54. C. Persson and A. Zunger, Phys. Rev. Lett. 91, 266401(2003).
55. J.G. Albornoz, S.M. Wasim and C. Rincon, Crystal Res. Tech. 34, 1191(1999); J.H. Schon, V. Alberts and E. Bucher, Semicond. Sci. Tech. 14, 657(1999).
56. F. Greuter and G. Blatter, Semicond. Sci. Tech. 5, 111(1990).
57. W. Kraak et al., Semicond. Sci. Tech. 51–52, 329(1996).
58. M.J. Romero et al., Prog. Photovoltaics Res. Appl. 10, 445(2002).
59. I. Visoly-Fisher et al., Adv. Mater. 16, 879(2004); Appl. Phys. Lett. 82, 1(2003).
60. O. Vigil-Galan et al., J. Appl. Phys. 90, 3427(2001).
61. T.P. Thorpe, Jr., A.L. Fahrenbruch and R.H. Bube, J. Appl. Phys. 60, 3622(1986).
62. J.W. Seto, J. Appl. Phys. 46, 5247(1975).
63. P. Chattopadhyay, Semicond. Sci. Tech. 10, 1099(1995).
64. P. Chattopadhyay, J. Phys. Chem. Sol. 56, 189(1995).
65. A. Fedotov et al., Solid State Phenom. 67–68, 15(1999).
66. V.N. Abakumov, V.I. Perel and I.N. Yassievich, in **Nonradiative Recombination in Semiconductors**, eds. V.N. Abakumov and A.A. Maradudin, Vol. 33, North-Holland, Amsterdam, 1991 .
67. S. Smith et al., Appl. Phys. Lett. 85, 3854(2004).
68. T.A. Wagner et al., Solid State Phenom. 80, 95(2001).
69. Y. Yan, M.M. Al-Jassim and K.M. Jones, J. Appl. Phys. 94, 2976(2003).
70. A. Antonelli, J.F. Justo and A. Fazzio, Phys. Rev. B60, 4711(1999).
71. M. Kim et al., Phys. Rev. Lett. 86, 4056(2001).
72. www.veeco.com/appnotes/AN79_ElecChar_RevA0.pdf
73. M.A. Schofield et al., Phys. Rev. Lett. 92, 195502(2004).
74. H. Hilgenkamp and J. Mannhart, Rev. Mod. Phys. 74, 485(2002); A. Gurevich and E.A. Pashitskii, Phys. Rev. B57, 03878(1998).
75. S.J. Pennycook et al., J. Eur. Ceram. Soc. 19, 2211(1999).
76. X.A. Cao et al., GE Global Research, Technical Report No. 2002GRC206, GE, Niskayuna, NY.
77. D. Cherns, S.J. Henley and F.A. Ponce, Appl. Phys. Lett. 78, 2691(2001).
78. J.Y. Shi et al., Appl. Phys. Lett. 80, 2293(2002).
79. M. Albrecht et al., J. Appl. Phys. 92, 2000(2002).
80. B.S. Simpkins et al., J. Appl. Phys. 94, 1448(2003).
81. D. Cherns and C.G. Jiao, Phys. Rev. Lett. 87, 205504(2001).
82. D.C. Look and J.R. Sizelove, Phys. Rev. Lett. 82, 1237(1999).
83. N.G. Weimann et al., J. Appl. Phys. 83, 3656(1998).
84. G. Koley and M.G. Spencer, Appl. Phys. Lett. 78, 2873(2001); J.W.P. Hsu et al., Appl. Phys. Lett. 81, 79(2002).

85. Y. Xin et al., Appl. Phys. Lett. 76, 466(2000).
86. I. Shalish et al., Phys. Rev. B61, 15573(2000).
87. I. Shalish et al., Phys. Rev. B64, 205313(2001).
88. I. Shalish et al., J. Appl. Phys. 89, 390(2001).
89. Y. Ivanov et al., Eur. Phys. J. Appl. Phys. 27, 371(2004).
90. S. Pizzini et al., Solid State Phenom. 95–96, 273(2004).
91. T. Arguirov et al., Solid State Phenom. 95–96, 289(2004); W.L. Ng et al., Nature 410, 192(2001).
92. J.A. del Alamo and R.M. Swanson, Solid State Electron. 30, 1127(1987); M.S. Tyagi and R. Van Overstraeten, Solid State Electr. 26, 577(1983).
93. E.S. Machlin, **An Introduction to Thermodynamics and Kinetics Relevant to Materials Science**, Revised Edition, Giro Press, New York, 1999.
94. G. Hanna et al., Thin Solid Film. 387, 71(2001).
95. C. Jiang et al., Appl. Phys. Lett. 84, 3477(2004); 85, 2625(2004).
96. J. Oila et al., Appl. Phys. Lett. 82, 1821(2003).
97. H.R. Moutinho et al., IEEE Photovoltaic Specialists Conference 2000, Conference Record of the 28th IEEE, p. 646.
98. A. Jasenek and U. Rau, J. Appl. Phys. 90, 650(2001); A. Jasenek et al., Appl. Phys. Lett. 79, 2922(2001); A. Jasenek et al., Appl. Phys. Lett. 82, 1410(2003).
99. D.L. Young and R.S. Crandall, Appl. Phys. Lett. 83, 2363(2003).
100. J.C. Bourgoin and N. De Angelis, Semicond. Sci. Technol. 16, 497(2001).
101. R. Krause et al., Phys. Rev. Lett. 65, 3329(1990).
102. O. Ueda, Mater. Sci. Eng. B20, 9(1993).
103. M. Uematsu and K. Wada, Appl. Phys. Lett. 58, 2015(1991); Appl. Phys. Lett. 60, 1612(1992).
104. J. Nissila et al., Phys. Rev. Lett. 82, 3276(1999).
105. R. Reitano et al., J. Appl. Phys. 74, 2850(1993).
106. W.G.J.H.M. van Sark, in **Thin Films and Nanostructures**, ed. M.H. Francombe, Academic Press, San Diego, 2002, p. 1.
107. R.E. Norberg et al., MRS Symp. Proc. 609, A 26.2.1(2000); P.A. Fedders et al., Phys. Rev. Lett. 85, 401(2000).
108. S. Hazra, A.R. Middya and S. Ray, J. Appl. Phys. 78, 581(1995).
109. S. Ramalingam et al., Appl. Phys. Lett. 78, 2685(2001).
110. S. Sriraman, E.S. Aydil and D. Maroudas, J. Appl. Phys. 92, 842(2002).
111. D.J. Leopold et al., MRS Symp. Proc. 715, A1.1.1(2002).
112. C. Banerjee, A. Sarkar and A.K. Barua, J. Phys. D: Appl. Phys. 35, 3060(2002).
113. Z. Yiping et al., Phys. Rev. Lett. 74, 558(1995).
114. M. Pippan and J. Oswald, SPIE 1985, 490(1993).
115. D. Han et al., J. Appl. Phys. 94, 2930(2003).
116. S. Yamasaki et al., J. Non-Cryst. Solid. 266–269, 529(2000).
117. T. Su et al., MRS Symp. Proc. 762, A12.4.1(2003); ibid. (2004).
118. N.H. Nickel and E.A. Schiff, Phys. Rev. B58, 1114(1998); C. Boehme et al., preprint submitted to Elsevier Science, 2004.
119. H. Fritzsche, in **Insulating and Semiconducting Glasses**, ed. P. Boolchand, World Scientific Press, Singapore, 2000, p. 653.
120. N.C. Greenham, X. Peng and A.P. Alivisatos, Phys. Rev. B54, 17628(1996).

Appendix

Figure A7.1 shows a schematic view down the core of a Lomer dislocation in the diamond cubic structure aligned with [001] vertical and [110] horizontal directions.

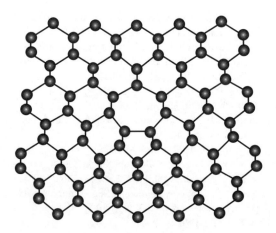

Figure A7.1. Modified with permission from M. Mostoller, M.F. Chisholm and T.S. Kaplan, Phys. Rev. Lett. 72, 1494(1994). Copyright 1994 by the American Physical Society.

Index